An Engineer's Life

One Man's Tale of Adventure

by
Alfred H. Bellows

2018

THE COVER

The cover colors—white lines and lettering on a dark blue background—are reminiscent of the blueprints that were common during my early years as I developed an interest in architecture. I often worked with blueprints during my summer job at Hamilton Standard. True blueprints always seemed to possess a degree of mystery and elegance not found in the later blue or black lines on white background. Therefore I chose its imitation for the cover. The line drawings on the cover are selected from 12 of the 33 patents issued in my name.

Front cover from upper left and spiraling clockwise: My drawing of power-jumper module from '802 (full patent numbers are found in *Appendix A*); simple automatic camera shutter from '194; microwave-powered automobile headlamp of '100; liquid-cooled extended scanner from '913 (and patent-drawn from my own freehand pencil drawing); and motor switching system for Polaroid SX-70 Land Camera's film door of '565.

Back cover upper row, left to right: Self-erecting camera bellows from '907; and first fully working Polaroid SX-70 of '123. Center row: Ultra-thin viewfinder with toric mirrors from '076; solar collector centralizer from '154; and residential circuit breaker desensitizer of '641. Lower row: Telephone line testing scheme for optical fiber-deployed system from '033; and one of many alternative camera configurations for the SX-70 camera from '770. (This last figure may be hidden by the ISBN label.)

For my Children and Grandchildren

I have often wished I had learned more about my parents' early lives and remembered more of what they did tell me. And as an engineer myself, I would like to have known more about the engineering careers and family lives of my two grandfathers.

> (Thus the original edition of this book was written and printed for my progeny.)

Dedicated to my Grandparents

"Were the offer made true, I would engage to run again, from beginning to end, the same career of life. All I would ask should be the privilege of the author, to correct, in a second edition, certain errors of the first."

Benjamin Franklin

An Engineer's Life
One Man's Tale of Adventure

Copyright December 2012 and August 2018
by Alfred H. Bellows

ISBN 9781728603599

CONTENTS

Preface	vii
Chronology	ix
Engineering: What is it like?	1
The Beginnings	2
Neighborhood Play	5
School Days	9
Youthful Experiences, Tales, and Anecdotes	14
Summer Jobs	28
MIT and Sigma Chi	39
Block Engineering	53
Outside Block and Polaroid: Hayes Street	54
The Polaroid Years	61
Outside Polaroid and ECA: The Apartment House	79
Electronics Corporation of America	85
Foster-Miller	95
Adult Experiences, Tales, and Anecdotes	103
The GTE Years	119
Osram Sylvania	140
Homes, Rentals, and Remodeling	149
Consulting	168
Dangerous and Stupid Doings	179
Inventions that Didn't ...	183
Retirement	189
Putting It All Together	193
Appendices:	
A. Patents, Publications, and Reports	199
B. Resume and Curriculum Vitae	205

PREFACE

This book was envisioned hundreds of times over the last decades of my engineering career. It would suddenly strike me, "I should write about this experience." *This experience* might be a sudden insight into a problem; the success of an uncertain experiment; a fruitful meeting with colleagues or clients; watching a product of my design going down the assembly line; the successful assembly of a complex instrument; or just an especially productive day of inventing.

This story is the life of only one engineer and the specific path he chose—a path of creativity, not analysis. This engineer was fortunate to have had a career that provided variety of focus and breadth of accomplishment. His life was very different from that of an engineer who practiced only one specialty, became an educator, or moved into the ranks of management. I cannot speak to those lives—just my own.

Being an engineer has always been a joy for me: the jobs, the projects, the challenges, the successes, the subtleties discovered, the people I've met, and the places I've been. I have even enjoyed the smells of a freshly-sharpened drafting pencil, a basement shop, or a well-greased machine. But none of those experiences have been known to the people who matter most to me. Therefore, I originally undertook this book to summarize some of that life in permanent form for my children, not just a jumbled collection of yellowed papers and boxes of old project remnants. I was abruptly awakened to its need following my first retirement party in 1997. After some of my GTE colleagues rose to speak of their work experiences with me, one of my sons remarked, "Wow, I learned more about Dad tonight than in decades of living with him!"

As I began to write of my jobs, it was necessary to write about incidents from my early life—after all, when did my technical life begin? At 24 going to my first day at a full-time job, at 18 beginning freshman year in engineering school, at 11 when first drawing and following a plan for building a house, or at six when first dissecting a broken appliance? Once I began addressing that period, it became necessary to fill in other pieces to establish context—who were the friends who helped shape my interest in building, or how did my family and teachers encourage or discourage my intrigue with science?

Thus this recollection is mostly about me, my professional life, the technical aspects of my home life, personal experiences, some anecdotes, significant people who have passed through my life, and occasional descriptions of family matters. The *I-*, *me-*, *my*-count in this book must be enormous. Sorry, but what were my choices? "The author designed the gadget"? "The gadget was designed"? "We designed the gadget"? "He designed the gadget"? Those are so awkward, unclear, inaccurate, or remote. After all, this book is about my projects, my work, my life, my accomplishments. I have tried always to give credit when it was properly *we*, *us*, and *our*. There are many places where, for variety and relief, I just omitted the personal pronoun: "Designed the gadget" despite the grammatical defect. Admittedly, there is a fine line between being boastful and being thorough, and hopefully this work is only the latter.

These stories were wrought mostly from memory. As such, there are undoubtedly unintentional errors, hopefully small ones. Although largely from memory, there were some incidents for which I looked up accurate data.

This book began to truly take shape in the winter of 2011 when I was approached by Christopher Bonanos, the author of *Instant: The Story of Polaroid* (Princeton Architectural Press, Sept. 2012) looking for information about the early days of Polaroid's SX-70 development. Not only did he interview me by phone, but he wanted to see the description I had written. When searching my computer, *the* description turned out to be *those*, which I collated and consolidated into a satisfying whole. The experience inspired me to shape-up the other false starts, and ultimately to write about unwritten periods. As the project progressed, it not only expanded, but I came to enjoy the recalling and writing more than ever expected. I hope readers also find it interesting and sometimes entertaining. (Challenge: Read the section about *logarithms* in the *School Days* chapter without chuckling.)

Many helpful, intelligent, interesting, delightful, and witty people walked through my life as both an engineer and a citizen. I wish I could name them all to honor their part in making my life a happy and useful one. Alas, it seemed best for the format of this book to leave most of them unidentified and for those few who are identified to do so simply. The latter group comprises persons who might be widely known or needed to be referenced repeatedly.

This book was originally written in 2011 and '12, named *Reflections: My Life as an Engineer*, and bound by me in a quantity of about ten. In the years since, several people have enjoyed it and encouraged me to make it available to aspiring engineers. Thus this revised and shortened edition renamed *An Engineer's Life: One Man's Tale of Adventure*.

Chronology of Key Points in my Life

1940 January: I was born in Rochester, New York, but the family soon moved to Charlotte, North Carolina where my father opened a private medical practice.

1942 Summer: One of my earliest memories at 2½ was watching a group of people using sticks to agitate concrete being poured into a wood form when Dad made a swimming pool for us. *Attracted to construction at an early age!*

1944 October: Moved from the rented house to a colonial house on 3 acres in a semi-rural area of Park Road.

1944 October: Met my long time friend Gene Carpenter. We were tight friends for about a decade and spent many hours developing roads for toy cars, building houses and tree houses, cutting grass, tending chickens, *etc.*

1946 – 1950: Together, my buddy, Gene and I built about six "houses"—five in the woods behind his home and one in our yard.

1947 – '48: Given a Donald Duck camera: a box camera, which was a delight.

1950 June: Went to Camp Sea Gull the first of three years for four weeks each.

1950 September: Entered fifth grade at Park Road Elementary School, a new school that I had closely watched being built. Elected to student council.

1951 Spring, Summer, Fall: Designed and built *My House*, a 7 by 7.5-foot house following standard building practices wherever possible.

1951 Christmas: Given a Kodak Pony 135 camera, a true joy to use and master.

1953 Summer: Took over tending our flocks of chickens and ducks, and selling the eggs to Mother. Got our first TV set, a B&W console unit, probably 11-in.

1954 September: Attended a private school for ninth grade only. Joined the photography project, became the school photographer using a Speed Graphic press camera, and developed film and 8x10 glossy prints in their darkroom.

1956 February: Got drivers license. License listed me as 6 ft 2 in., 145 lb.

1956 July: First auto accident while driving my date and two other couples to an amusement park. The car was totaled, but I put it back together in the driveway with a junkyard front end for $30.89.

1957 October 4: Sputnik launched by USSR. Along with 3 or 4 others, I was invited to the principal's office after school to be interviewed by *The Charlotte Observer* regarding the significance of the Russian space program launch.

1958 Winter-Spring: Built a Michaelson Interferometer as a science fair project. Fabricated and assembled all parts including silver mirror flats. Placed well enough to go to the N.C. State Science Fair at Duke University.

1958 June: Graduated from high school. First paying job: a day camp counselor. Drove one of the school busses (18 years old, no special license!).

1958 September: Rush week at MIT. Pledged Sigma Chi and moved in. Began classes a week later. Major: physics, then later mechanical engineering.

1959 Summer: First technical job—Duke Power Company testing insulation on high voltage transformers and circuit breakers (up to 100,000 volts).

1960 Summer: Another job at Duke Power testing generator insulation.

1961 Spring: Designed a variable speed drive, a class project.

1961 Spring: Working with two other students, we designed and built a "broom balancing" machine, the *Servo-Controlled Inverted Pendulum*.

1961 Summer: First engineering job: Raybestos Div. in Stratford, Conn. At midsummer got a $5 raise to $100 per week. Designed factory equipment.

1961 Fall: Designed a Pedal Control for an automobile, a dreamed-up project for the handicapped. Won an award, and later entered into a welding contest.

1961 School Year: Selected the thesis topic of designing an optical model for use with ultrasound neurosurgery research for doctors at Massachusetts General Hospital. A hand-built optical bench was used by thesis advisor Prof. Mann to pitch the case for accepting me into graduate school. Many hours were spent in the shop during spring term making all the parts. A press conference led to an article in *The Boston Globe*.

1962 June: Graduated from MIT with a BS degree in Mechanical Engineering.

1962 Summer: Worked at Hamilton Standard, Windsor Locks, Conn. Designed a fuel-enrichment adaptor for the jet engine fuel control for Boeing 707 planes.

1962 September: Entered MIT Graduate School for a Master of Science degree. Employed as a research assistant for which I was paid to design and build my thesis—a Braille reading machine.

1963 January: Bought a Miranda DR camera, an all-mechanical 35 mm single lens reflex. Purchased a second body without lens within a year.

1963 Summer: Worked at Scott Paper Company's research center, Philadelphia. Designed, built, and tested a novel method of "creping" paper.

1964 January: First trip west of the Mississippi. Flew to San Francisco for a job interview and then spent four days enjoying and photographing Yosemite Valley virtually alone in three feet of snow.

1964 April: Purchased a used 1961 Volkswagen beetle specifically for my planned summer trip around the country. Light blue, cloth sunroof, $1,200.

1964 June: Awarded a Master of Science degree.

1964 Summer: Spent the summer on a huge loop around the country, living out of my VW. The trip took 75 days on the road, 65 sequential nights sleeping on the ground or in a jungle hammock, covered 13,124 miles, and cost $511.

1964 September: Began work at Block Engineering, Cambridge, Mass. Engaged in minor work on an infrared interferometer-spectrometer. Soon resigned.

1964 December 14: Began working for Polaroid Corporation in the camera design division in Cambridge, Massachusetts, a few blocks behind MIT.

1965 May: Bought a Calumet 4x5 view camera. Bought and rebuilt a used 4x5 enlarger and installed a darkroom in the kitchen.

1965: First invention acknowledged by an issued US Patent. Invented a hidden bistable hinge mechanism to cycle a camera's bellows open and closed. Pull—it snaps open to take pictures; push—it snaps closed to carry. (See below.)

1965 December 14: Met Dr. Edwin H. Land when my boss took me to his office to demonstrate six weeks of progress on a working scanning camera, a forerunner to the Polaroid SX-70 Land Camera.

1966 March: The first of my 33 issued patents was filed. Patent No. 3,418,907 "Photographic Camera Erecting System," (the self-erecting bellows!).

1966 June: Sent by Polaroid to Ansel Adams's Yosemite Photography Workshop for two weeks. Fabulous experience.

1967 June: During several phone calls from Dr. Land who was in Palo Alto at the time, he and I reinvented the scanning camera into the odd-angle folding camera ultimately sold as the SX-70. Patent No. 3,683,770.

1968 July: Made an offer on a 4-family apartment house in Cambridge, Mass. Passed papers on September 19. Ultimately spent over 2,000 hours repairing and remodeling the building and its apartments.

1969 August: Got married—the first time.

1971 March 15: Began work at ECA as Chief Mechanical Engineer, Cambridge.

1971 Summer-Fall: Revived a languishing project into a complete water-cooled scanner for monitoring the interior flame of an electric utility boiler.

1972 August: Our daughter was born.

1973 March 29: End of the Viet Nam war, a dark period among many of my generation.

1973 Summer: Purchased a 132-year-old house in Belmont, Mass. Continued renting it for a year, then moved in.

1973 - 1974: Designed the optical system, hardware, plastic molding details, and assembly jigs for world's first infrared LED-based photoelectric scanner. Managed a "record-breaking" race into production.

1974: Within a three week period an electronic engineer colleague and I designed and produced 100 prototype units of a new product by utilizing available components in ECA's stash and only two custom but simple parts.

1975 March: Promoted from Chief ME to Manager of Manufacture at ECA.

1975 Summer: Revamped a factory's awkward production schedule into a semi-continuous plan that reduced production-line labor by nearly 20%.

1975 September 29: Fired from ECA.

1975 October: My wife with our daughter moved out.

1975 November 10: Started work at Foster-Miller Associates, Waltham, Mass.

1975 December: First time of many crawling through a coal mine.

1976 September: Bought a 12-unit apartment house in Waltham.

1977: Designed and built articulating model of a novel coal mine roof support.

1978: Divorce settled. Thankfully it included joint custody of my daughter.

1978 May-June: Six weeks in a coal mine testing flexible-shaft roof driller.

1978 September: Sold the apartment house in Waltham.

1978 September: Married my second wife. Two stepsons joined my family.

1978 December 26: Started work at GTE Laboratories Incorporated, Waltham.

1979 – 1982: Designed modular circuit breaker, its tooling, assembly, and tests. Also supported other breaker and high-voltage projects for Sylvania and Zinsco divisions. Those divisions were abruptly sold in 1982.

1979 August: Sold the Belmont house and purchased a 6-bedroom, 11-room house on about 1.1 acres in Wayland, Mass.

1980: Began a ten-year consultancy to Avant Inc., designing ID cameras.

1981 June: Our son was born.

1983: Conceptualized, designed, built, and tested Get Away Special payload for flight on Space Shuttle to measure zero-gravity effects on 175 watt arc lamps.

1983 July: Spent the entire month on a western journey with all four children in a 21-foot Winnebago. Flew to Denver, then drove to Rocky Mountain NP, Grand Teton, Yellowstone, through Idaho, Salt Lake City, Winnemucca, Reno, Lake Tahoe, Yosemite, San Francisco, Los Angeles (Magic Mountain amusement park), San Diego, Tijuana, Las Vegas, Zion Canyon, Grand Canyon, Four-corners, Durango, Silverton, Black Canyon of the Gunnison, and past Pikes Peak back to Denver.

1984 February 3: Watched our launch of STS-41B (STS-11), Shuttle Challenger, which orbited for 8 days and successfully ran all three of our experiments.

1984: Temporary assignment to GTE's Government Systems Division to design surge-arrestor for up to 600,000 amperes for Minuteman missile bunkers.

1985 - 1992: Planned, designed, and managed another Get Away Special project for testing the effect of gravity-induced convection on quality of crystal growth in gallium arsenide, a more static-resistant electronic crystal than silicon.

1986 Summer-Fall: Added a 12 by 18 foot extension to our home and remodeled the kitchen. All work was done by me except for pouring the foundation, electrical wiring, oak flooring, and custom cabinets.

1987 April: Over a weekend, learned to program a 3-inch square microcomputer in BASIC to control and collect data from the GaAs crystal growth payload. First of many Tattletale-controlled projects I programmed.

1987 April: Sold the Cambridge apartment house after 19 years.

1989 – 1991: Served the Town on the Road Construction Committee. Chair one year.

1991 – 2001: Served the Town on the Finance Committee. Chair two years.

1991 June: Orbited the first flight of the "GaAs GAS" crystal growth experiment aboard Space Shuttle Columbia, STS-40, as GAS Payload No. G-052.

1992 March: Second flight of the GaAs GAS crystal growth experiment aboard Space Shuttle Atlantis, STS-45, as GAS Payload No. G-229.

1992: Managed a three-person project to develop a fiber to the home connection for telephone service—arguably an early version of FiOS.

1994 – 1997: Conceptualized, designed, built, programmed, and implemented field testing of the *Back Watch*, a device that monitored up to 24 auto-sized batteries in back-up service at eleven outdoor telephone equipment sites.

1995: Became an expert witness for a GTE telephone division in defense of liability for a serious electrical injury caused through negligence by the local power company on whose utility poles GTE was a "tenant."

1995 – 97: NASA contracted ABell Engineering to design and build a Resistance Measuring Device (RMD) for materials research within the space shuttle.

1997 September: Retired from GTE Labs after nearly 19 years and scores of fascinating projects.

1997 November: Began work at Osram Sylvania on a 4-day week basis.

1998 – 2000: Designed hardware and instruments for several Sylvania projects including molded arc tubes, accelerated filament testing, infrared video scanning, and vacuum processing of exotic alloys.

1998 June: Got a call from Stroock & Stroock & Levan, a NYC law firm, asking me to be an expert witness for them and their client, Fuji Photo Film Co. Although described as a relatively brief project, it ended up as 2,203 hours over 11 years with 65 hours of deposition and 58 hours on a witness stand during four appearances at the International Trade Commission and four in US District Courts in New Jersey.

2000 July: Joined the Osram Opto Semiconductor group based in Regensburg, Germany although my office remained at Osram Sylvania in Beverly.

2001: Provided electrical configuration for efficient powering of LED lighting at the Jefferson Memorial.

2005 January: Retired from Osram Sylvania at age 65.01 after seven years.

2006 December: Purchased a house in Osterville, Cape Cod, Massachusetts.

2007 March: Bought a condo in Naples, Florida.

2008 February: Elected to our Condo Board in Naples, beginning a ten-or-more year tenure as president and workaholic. Major projects included managing the installation of an entrance gate; evaluating and overseeing geothermal heat for the swimming pool; introducing "pooling" for the reserve fund; and upgrading the parking pads to paver blocks.

2008 June: Moved from Wayland to Osterville, Mass.

2011 – 2012: Wrote, designed, formatted, and bound *Reflections: My Life as an Engineer*, a 384-page book including introductory pages and four appendices. This book was for my children and only about ten copies were published.

2011 Spring: Purchased, built, and began sailing *Salty III*, a Soling class radio-controlled model yacht.

2015 May: Bought a condominium in Newbury, Mass. to which we moved in two steps: June and December.

2017 August 21: Experienced my first total eclipse of the sun with my son in Mokane, Mo. He gave me the trip for Christmas along with two baseball games: Royals and Cardinals.

2018 July: Performed a marriage for two delightful young friends. My second experience as an officiant.

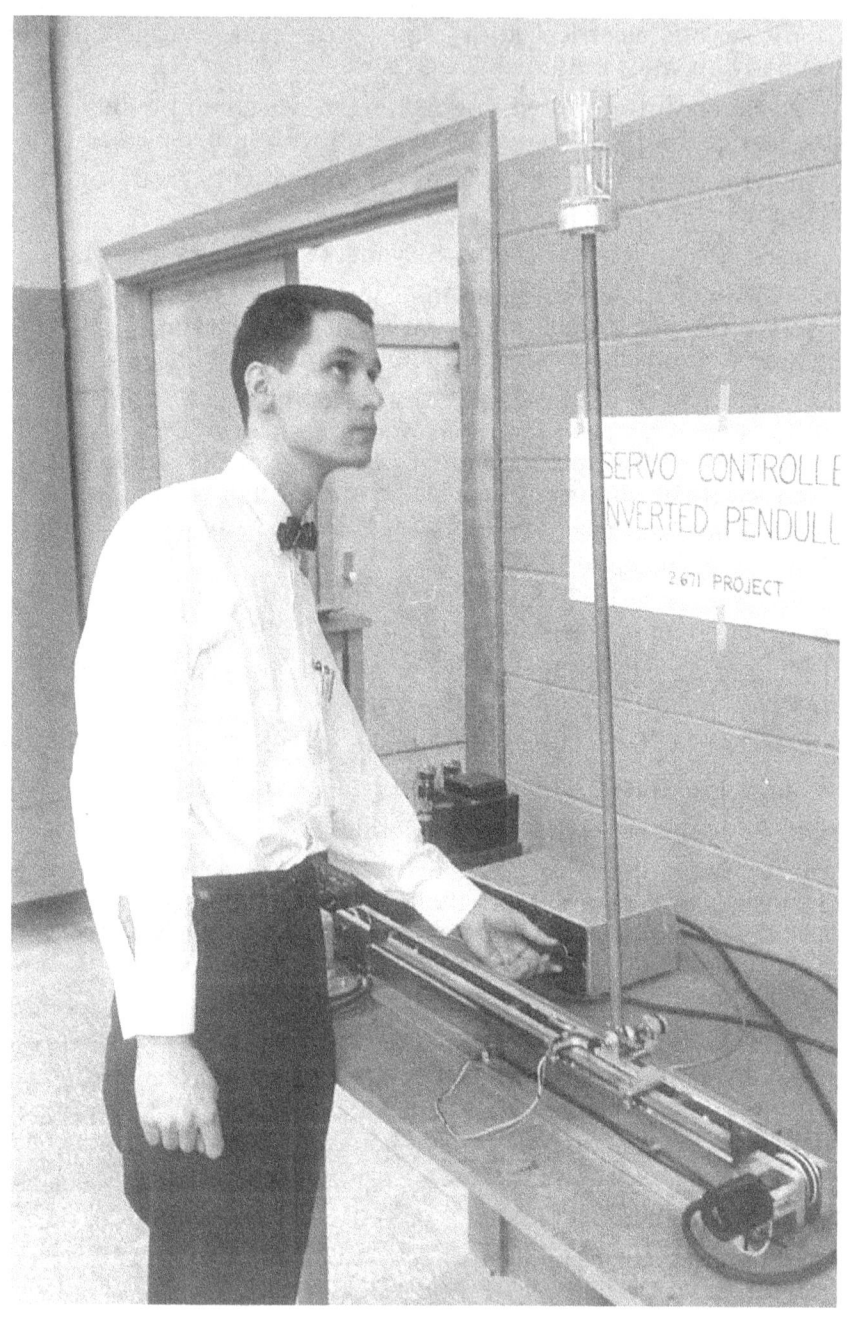

The author demonstrating the Servo-Controlled Inverted Pendulum, a junior year project, 1961.

ENGINEERING. What is it like?

It is physics—The logic fascinates me.
It is math—I like mathematics; and am good at it.
It is building—Creativity is fulfilling—A kind of art; start with nothing, end up with something.
It is rigor—Do it sloppily, and it falls apart; do it well, and it is beauty.
It is variety—One day mechanics; the next electronics; another day optics; next week explaining and defending a gizmo before a group of cynics.
It is economics—If the cost is too high, no one will buy it.
It is design—Try this way; try another way; which is better?
It is analysis—Does this work? is it strong enough? is it light enough? will it be beautiful?
It is design—Do that "trying" thing fifty more times and you might be finished.
It is diversity—research, synthesis, analysis, introspection, advocacy.
It is documenting—lab notebooks, photography, taking data, sketching, writing reports.
It is nature—The more natural a design the better; if you have to work the design too hard, it is probably a weak approach.
It is a paycheck—Can be a good one, too; and in exchange for having fun.

It was the most fantastic career I could ever have imagined. Could there have been a better one than engineering? Not for me.

From mechanics to electronics; from optics to microcomputers; from product development to conducting seminars; from quiet concentration to traveling the countryside; from sophisticated cameras to mundane brackets; from microelectronic chips to gimbaled structures; from underground mining machinery to computerized Space Shuttle experiments; from defining details to managing whole projects; from surgical instruments to exploring factories; from bosses to protégées; from service to the community to expert witness in federal court; and many more—they all had their challenges, frustrations, intrigues, compromises, and debuggings.

Moreover, they all had bosses, colleagues, staff, technicians, customers, marketing, *et al.* to negotiate with and satisfy.

So, how was I so fortunate? Please read onward.

THE BEGINNINGS

It all started in January 1940 when I was born in western New York State and brought home to meet my older brother. Within a few weeks, my parents moved us to Charlotte, North Carolina where I spent my entire youth.

My parents were both born in 1903 before airplanes and radios. My father grew up in central New York State and my mother in South Carolina. They met each other at Duke Medical School where Dad was an intern, and Mother was a physiotherapist. Soon after their wedding in 1932, Dad began a residency in Rochester, New York. Hence my birthplace.

My only sibling is William, three years older than I. In our early years, Bill was far more gregarious and much more disciplined than I. He loved sports and took school seriously. My interests became more technical than his, and our friends and activities rarely overlapped, although we enjoyed vacation adventures together and worked together on many projects around the house when we were teens.

• • •

Grandparents – Both of my grandfathers were engineers, one by training, the other by intuition. Surely some of my interest in engineering came through those genes and undoubtedly some came from years of following my mother's father around his house and yard as he solved many problems and invented many gadgets.

I hardly knew my paternal grandparents who lived some 800 miles away and died before I was a teen. Grampa Bellows was a civil engineer who had degrees from Brown University and spent most of his career working for the State of New York. During my father's youth, Grampa was designing locks for the Barge Canal, the expansion project of the Erie Canal. In the 1920s, he was a design engineer for the Holland Tunnel between Manhattan and New Jersey. Gramma grew up in Connecticut where her father, a Civil War veteran, owned and operated the general store.

• • •

I have many memories of my maternal grandparents. My grandfather, Alfred Hammond, for whom I am named, was born in South Carolina across the river from Augusta. He was known as *Alf*, but I only knew him by his later nickname, *Old Man*. As a young man, he built and operated a cottonseed oil mill on the family's millpond. He left the family home soon after marrying at 27 and settled in Columbia, S.C., where he worked for most of his professional life at Swift & Company Oil Mill, one of their cottonseed oil plants. For most, if not all of his years, he was the superintendent of that plant. He was awarded one patent that I know of.

My earliest memories of him were as a white haired, bald, old man with a rare but engaging smile and lots of patience with us fractious kids. As an example of his patience and energy, when he heard that I wanted a pile of sawdust to play in, Old Man constructed a chest and filled it with sawdust as a Christmas present when I was nearly four and he nearly 71. The chest was handmade by him of wood and "tin," as he called galvanized sheet metal. (When ten or twelve, I stained it and still have it today.) The body is fashioned of various widths of 1-inch boards nailed together in a cleverly-rigid design and measures about 1 by 1 by 2 feet. The top is a wood frame with tin panel tacked in place. It is secured with purchased hinges and hasp, but is propped open with an over-center lid support contrived by Old Man of heavy sheet tin and pivot pins made from riveted nails. Each end has a rope handle with handmade tin cleats nailing it in place.

Old Man welcomed me into his shop where I loved to watch him work at his messy workbench. He was very clever and could make almost anything. His drill press was handmade of wood and a manufactured crank-operated breast brace. Some of his screwdrivers were fabricated of uncoiled spring steel, rehardened, and outfitted with leather handles. He made his own wheelbarrows, constructing the oversized wheel from 1-inch boards joined with wood braces at 90 degrees on opposite sides, hand-sawn round, and rimmed with pieces of V-belt rubber tacked in place. He taught me that large wheels roll over bumps more easily.

My grandmother was called *Lady* by us. She was mostly a stay-at-home mother, but at some period she worked for the police department as a juvenile officer where she was issued a pistol. We always thought it cool that Lady once had a pistol. Lady was a warm and caring grandmother who gave me unlimited attention as I needed it except when a baseball game was on the radio. She could not be distracted from those games.

As much as I adored Lady, I mostly spent time with Old Man whenever we visited their home. If there were noises from his shop below, I would go to see what he was doing. I followed him around the yard as he puttered—cutting grass, trimming bushes, sawing a board, setting up the ladder, tending his quarter-acre garden, or some of his many other projects. I learned a lot of "engineering" by watching him.

My Parents – Mother was at home throughout my youth and spent many hours with Bill and me. We had an adequate collection of toys, and she read to us a lot, both together and individually. She often knitted while reading. She took us on picnics to the nearby park and on many walks in the woods. She taught us a lot about nature: trees, flowers, insects, animals, reproduction, the stars, and a very detailed and accurate account of the solar system, the seasons, and the phases of the moon. She also taught us how to take off a wet T-shirt. She loved knitting, embroidery, and sewing: making many of her own clothes and a few shirts for the rest of us. During the car and gas shortages of World War II, she took me on many excursions on the back of her bicycle. She could wield a hammer and saw and helped and encouraged me with many of my craft or construction projects over the years. She undoubtedly had an influence upon my becoming an engineer.

During my early years, Dad was the only neurosurgeon in all of North and South Carolina and worked long hours so he didn't spend much time with Bill and me. He did, however, fit in the time to make some toys for us—a jungle gym, falling blocks, swings, sandbox, crutches. And he built the forms for and made a concrete above-ground swimming pool for us, surely a time-consuming project that he did for our benefit.

That's me at 16. Sophomore photo taken for the 1956 Myers Park *Mustang*, our high school yearbook.

Dad was a bit of a curmudgeon. He knew that about himself and deliberately chose his arcane medical specialty so that he would be recommended by doctors for his competence rather than by former patients for his personality. He could hold grudges for years and showed little respect for the needs and accomplishments of others including his own family. More than once, I was the brunt of his anger over trivialities, but eventually our separations were dissipated if not truly resolved.

As I got older and could engage with Dad at a more adult level, we developed a much closer relationship. He was keenly interested in science and numbers and as my reading began to include things like *Scientific American* and George Gamow's *One, Two, Three, Infinity*, we got into some interesting discussions.

He liked undertaking projects around the house, and I often watched him at work when I was little and worked with him as I got older. I learned many skills from him and had many enjoyable moments with him during those periods.

NEIGHBORHOOD PLAY

Although I didn't know the word *engineer*, I knew from an early age that engineering was my destiny. Observing, planning, figuring things out, and building were favorite pastimes. Some of my early play experiences undoubtedly influenced the engineering life that I took up. A few examples follow.

Shortly after we moved to Park Road when I was four and a half, I met a nearby playmate. Gene and I became constant companions for many years, playing in the attic, the basement, the chicken yard, the woods, the cow pasture, and all around the neighborhood. Gene and I shared a great interest in building things and we built many houses together in our backyards.

I had four other regular playmates in those early years—Walter, James, Sven, and Wayne. Walter's father was a private pilot at Morris Field and often came home with World War II surplus airplane parts for us to play with. Some were used in construction projects and others were used to satisfy my curiosity about mechanisms such as the viewing portion of a bombsight with its reticle, pantograph, and optical system.

James had a large flat front yard ideal for ball games, and in our teens, we worked on cars together. Sven had a great back yard, and Wayne lived on a small farm with tractors, hay balers, cows, and a smelly creek.

Our house was a two-story colonial on a three-acre lot. The neighborhood stopped abruptly four houses beyond ours and became farmland, mostly dairy. There was much to explore. The dirt roads near and behind us were home to several children of our ages with whom we occasionally played. The area behind houses on the other side of our road gave way to woods, a dairy farm, and a lake in the cow pasture.

After World War II ended, signs of growth emerged. Periodically, a new house would be built nearby, and I explored every one on a regular basis, absorbing everything I could about house construction up through the stages before locked doors were installed. In 1949, the farm beyond us was developed for a 12-classroom elementary school and two rows of small, single-story homes. I watched the construction of the school and the houses, and later met the architect of the school who showed and explained its blueprints to me.

Our yard included several outbuildings: a two-car garage with an attached room behind it, a smokehouse, a chicken house, a construction shack, and a barn. The room behind the garage had a potbelly stove that we used for heating during Cub Scout meetings, and a front porch about 8 by 10 feet where I rebuilt carburetors, engines, and other machinery that was too messy or big for the basement. My mother repaired the fence at the chicken house and started keeping chickens for fresh eggs and food. In time, I became interested in the chicken-raising and eventually took over from her and got a Boy Scout merit badge for it. The barn was about 30 feet square with an attic. We played there a lot, and I have a glorious memory of lying on a pile of straw in the attic doorway with the sun streaming in upon me.

Although I didn't care much for team sports, I did like swimming, diving, sailing, motor boating, canoeing, archery, riflery, and hiking. In tenth grade I bought hockey skates and learned to skate at the newly-built Coliseum where I also wangled a brief job as "human Zamboni" when the city acquired a hockey team—the Charlotte Clippers. I took up water skiing in my teens, snowshoeing in my twenties, and alpine skiing in my forties. In high school PhysEd class, the coach considered me to be a goldbrick, but usually lessened my punishments because of my outstanding distance in the shot put, my arms-only rope climbing, and my ability to figure out the combination to dozens of otherwise-useless locks he had collected in a box.

Much of my playtime with a friend or friends of the moment was *exploring.* We rarely wandered more than a mile, but we knew our neighborhood: the creeks, the woods, the new construction projects, the farms, and the pastures. It was in the pastures that we found rabbit tobacco, a stalk plant with small, silvery leaves that could be smoked after drying. I made corncob pipes for each of us using small-diameter bamboo sections for the stem and popcorn cobs for the bowl. That stuff couldn't be inhaled, and we rarely smoked, but it seemed so daring to do so at all! And it was in all those places that I learned to spot and try to avoid poison ivy, since I caught it big time.

• • •

Perhaps my most memorable single form of play was house-building, mostly with Gene. I remember assembling a complex of boxes into a multiroom house before turning five. In the rear portion of Gene's yard there was about half an acre of pine woods where we first built a flimsy house between trees. I was probably six then. The next summer, we used rusting metal signs, boards, and tarpaper to build a house about four feet square that utilized two trees at the front corners for stability. This house had a hinged door—leather hinges held with nails—and a shed roof with a skylight of wire-reinforced glass. I had been given my own 13 oz. hammer for my seventh birthday, and both of us were pretty good at using hammer and nails by that summer.

While not banging nails, we cleared extensive roads throughout the woods for our wagons, tricycles, and bicycles.

Our biggest project together was a tree house started when I was about nine. We found a triangle of trees spaced about 3, 6, and 6 feet. We nailed boards between the narrowly-spaced trees to form a ladder, then built a framework around all three trees and covered that with flooring about 6 feet above the ground, well above our heads. Then we added walls around the trees, left a doorway at the ladder end, and covered it with a roof. The next summer we added a second platform below the first one and enclosed it to form a two-story tree house. I remember sitting on the first floor of this house before it had been enclosed while listening to a crystal radio that I built from some crude instructions and a minimal set of parts.

There were two or three other houses: one comprised hundreds of pine boughs hanging from dozens of wood slats nailed to six trees; and another was attached to the outside wall of the construction shack.

Shortly after my eleventh birthday, Dad decided to tear down the barn and use some of the wood to build a shed more suited to storing firewood. I was delighted at the prospect of having an abundant source of wood to build a *real* house—*my* house.

"My house," the house I built in 1951 at age 11. (Photographed several years later in January 1955 by me using my Kodak Pony 135 camera.)

By that time I had watched many houses being built, knew how to use tools, and was ready to build one *correctly*. Also, by this time I was interested in architecture, had designed several structures, built a cardboard model of a new chicken house, and visited an architect's office where I traced the plans of a nearby church. It was now time to draw plans for my first real house. The studs were laid out to accommodate the window sashes and door on-hand. The frontal width was 7.5 feet, the front to back depth 7 feet.

[**NOTE:** *The following and other inset paragraphs with smaller font found scattered throughout this book and introduced by a brief heading set in uppercase bold might be of special interest to technologists and historians, but many readers may prefer to skip over them as being too arcane.*]

CONSTRUCTION - At each corner, I dug away the topsoil, made wood forms, mixed concrete, and poured footings for four brick piers. Then, using mortar and some of the many bricks scattered around our yard, I built four columns two bricks square and 3 to 8 bricks high as needed to get the tops level. Atop the columns a frame of 2x8 band joists and joists 16 inches on centers was built. The joists were covered with barn-siding boards, 1x8s to form the floor deck. By the time walls were to be erected, it became clear that Dad was harboring most of the wood for his shed, so, reluctantly, I had to compromise on details: the bottom stud plate was omitted, studs were toe-nailed directly to the flooring, and the side walls had fewer studs than the standard of 16 inches. Also, planks for the siding clapboards had become scarce, so I used white-painted shiplap from a neighbor's demolished garage for the front and back, while the two sides were finished with rough-cut barn boards that I stained brown.

One window was installed in each of the front and back walls. The windows were single sashes about 20 inches square that had been found in the shack. The stud layout anticipated the sashes so that a 2x6 sloping sill and a 2x4 header framed a perfect space for the removable sash without need for a separate frame. The door was a 5-panel door, and I mounted it with proper hinges. Stops for both the sashes and door came from 1x3 fence pickets, nailed to the studs, and served as the exterior casing by being slightly proud of the shiplap clapboards.

The roof was to be gabled with the ridge running front to back. But when time for the roof came, I was stymied by both a shortage of 2x6s and my inability to assemble an unstable structure alone so far above my head. My grandfather, Old Man offered to help, and together we got it done. However, his solution departed substantially from the standard practice of a ridgepole with rafters butted to it. He chose to omit the ridgepole and made individual angled and rigidly-lapped rafter assemblies of 1x8s with a bolted-on cleat at the crows bite for nailing to the top stud plate. The roof was then planked and covered with green roll roofing—probably the only item other than nails purchased new for the entire house. Despite my regret for not following tradition, it was a good lesson about his ingenuity.

I remember taking a picture of the completed house on Christmas Day of 1951 with my new Kodak Pony 135 camera as Dad was teaching me to do f/-stops and shutter speeds. I was a month short of 12 years old.

I hardly ever played in that house *per se*, but as we got older some of my friends and I would occasionally have sleepovers, but only two or three at a time in that tiny space.

SCHOOL DAYS
1946 to 1958 – Ages 6 to 18

There are many memories from my school days that I cherish: beloved teachers, friends, games, frustration with sports, extracurricular activities, and of course the academics. A few stand out as affecting my technical development or having some other impact on my life.

When I went to first grade, I was looking forward to it and surprised my parents with my independence and how much I liked it. I had not learned to read yet but had learned to add single digits and was proud of being ahead of most other students—at least for a few weeks. Miss Neely assigned and rotated "important" responsibilities to us, like feeding the fish, opening the huge windows with a hooked stick, turning on the lights if needed, and clapping the blackboard erasers outside.

In second grade, Miss Wagoner organized the designing and making of silkscreened curtains for the windows. It was an involved process that required three overlaid stencils of flower designs that we seven-year olds spent weeks drawing from live flowers and cutting out using sharp X-Acto knives. One stencil had two layers, and I was one of the few who could get the hang of cutting through just one layer and lifting the loose piece off the backing. I enjoyed every step of this project and remember the bolts of white cloth, the huge silkscreen frames, the attached stencils, the teacher pulling the long squeegee with paint puddle across the frame, and the advancing of the cloth between each imprint. That had to be repeated on subsequent days after each color of paint had dried and the stencil had been changed out. What memories.

It was also in second grade that I first observed that the spelling books were bound in a distinctly different way from most books which led to my interest in bookbinding and the making of my first book—a collection of six or eight Katzenjammer Kids comic books, signature sewn to a piece of cheesecloth, glued together, and then glued into a hard cover of cardboard and muslin.

Miss Wagoner took us for a tour of a bread factory. I remember watching a pancake-shaped piece of rolled-out dough going down a conveyor belt, passing under a short piece of ordinary chain link fence that was dragging on the belt, being rolled up into a cylinder by the friction from the dragging fence chain, and dropping into a loaf-pan. What a simple, unobvious use of a piece of fence. That moment may have started my engineering career.

My only memory from third grade was gazing endlessly at Sara Ruth's pigtails.

The most significant memory from fourth grade was an occasion when the teacher caught someone violating another's property rights—like stealing their pencil or other 'valuable' item. She was very indignant about such behavior, walked to the front of the room, looked up at the wall above the blackboard, spread her arms in despair, and read one of the Ten Commandments from a framed listing: "Thou shalt not steal." Then she turned back to face the class with a dark and hooded look and said, "You just *don't do that!*" while shaking her jowly face back and forth. It's an image that comes back to me whenever I am tempted to do something clearly wrong. Too bad that today's youth cannot receive the integrity and wisdom that Mrs. Houston imparted to me by that simple demonstration so many decades ago.

In fifth grade, I went to the new Park Road Elementary School about two blocks from home that I had attentively watched being built. Our teacher, Mrs. White had a brother in the Korean War, and periodically she read his letters to us, which really made that conflict real to me. At Christmas time, we painted a Santa Claus scene over the entire classroom window and had a photo of it printed in the newspaper—my first appearance in the news—and my first experience at seeing a big project being organized.

For both fifth and sixth grade, I was elected to be a patrol boy, which meant wearing a white belt with badge over my chest, standing on one of the sidewalks before and after school, and making sure students were orderly. Also, I was elected to student council and in sixth grade, elected as its president.

In seventh grade, I went to the much-larger Myers Park High School with about ten classes per grade instead of two. As a result, I was much less known and never again got involved in school government. This was also about the time that I first sensed or recognized myself to be an "out-of-it" kid. I didn't integrate easily with the 80% of the class that came from other elementary schools and tended to pal around with my old buddies, some of whom were a bit nerdy themselves. I didn't dress fashionably, play any sports, participate in student government, or join the smoking group. To whatever extent I was aware of my status, I didn't seem to care enough to try changing. Had I been at a different school, shorter, or both, I probably would have been bullied.

For ninth grade, I attended a private day school. I enjoyed that year and got caught up in English, my previously weakest subject, learned to study and think, and had some interesting and interested teachers. That year, the school purchased a Speed Graphic press camera, built a well-outfitted darkroom, and established a photography program. I signed up, learned a lot from a professional photographer who came in to instruct us once a week, and effectively became the school photographer. I supplied several sports pictures to *The Charlotte Observer* and many of the pictures used in the yearbook published that spring. I also felt that my fellow classmates had respect for my talents, not contempt for my eccentricity.

That year I researched and wrote a paper on Archimedes. Being a mathematician and engineer, he was a bit of a hero for me, and I thoroughly enjoyed doing that paper.

Alas, Country Day was only K through 9, so it was back to Myers Park in tenth grade. I was enjoying school and getting much better grades, including previously unheard of *As* in English. I took industrial arts that included a woodworking shop and mechanical drawing class that I thoroughly liked and did well in. I loved biology and developed a great respect for the teacher from whom I could learn almost anything with minimal study. The insect-collecting project required in her class initiated a more casual habit that continued throughout high school, and by the time I left for college, there were fifty or more insects including beautiful moths and butterflies pinned to the walls of my bedroom. I took chorus all three years of high school and was one of the few men who could reach the bass range, so that's what I sang although tenor was my range in college and beyond. We had to memorize the choral parts of the Messiah, the respective portions of which we sang at assembly every Christmas and Easter with the orchestra.

I got a new combination lock for my locker that year. It was a *Yale*, and I personalized it by filling black paint into the recessed Y and E leaving AL in white letters in the center of the dial, which always ended upright because of the final digit in the 6-17-0 combination!

I joined the Engineering Club and became president senior year.

I wanted to take every math course offered, couldn't fit them in, and took plane geometry and advanced algebra at Charlotte College in the evenings during the summers. I was expecting this to put me in good stead for engineering school, but almost all of my fellow freshmen at MIT had studied calculus, which I had hardly even heard of.

In the eleventh grade, I took chemistry, which I thoroughly enjoyed. I also took typing and am so grateful for taking the time to learn touch-typing, a skill I have used regularly ever since.

A research paper on the Panama Canal took me to the public library where I pulled out ancient copies of magazines, especially *Scientific American*, which had several major articles on its progress. It was fascinating research that revealed so much real-time discovery for me of the many detours that project undertook over its years of planning, progress, failures, restarting, and eventual completion. In the end, I brought all my notes together and wrote the paper. Ultimately, I had to type it with its many citations, an arduous task on a clunky typewriter, and eventually submitted it for an *A*. One of my best grades ever in English.

In twelfth grade my favorite classes were physics, architectural drawing, solid geometry, and trigonometry. I was also a chemistry aide during one period, and this was the year that I built and demonstrated a Michelson interferometer for Science Fair. I used a few of the chemistry-aide periods to learn about and make the full- and half-silvered mirrors. The interferometer got first prize at the city level Science Fair, so I went to Duke University for the State Fair. I didn't win there—perhaps because I couldn't seem to get the interference fringes to display well enough for one of the judges to see. (See *Michelson Interferometer* in the *Youthful Experiences, Tales, and Anecdotes* chapter.)

Physics was my best subject. I liked the teacher but he was not very good at physics. Fortunately, the Charlotte school system had purchased a collection of 150 half-hour movies made by Prof. White of UCLA that took us through the entire range of physics lessons. As with biology, I absorbed every word. (Usually, I ran the projector so I got pretty good at using that classic Bell & Howell machine, too.) Sometimes during discussion, the teacher would call on me to answer a question.

I particularly liked trigonometry, perhaps because of its relationship with surveying, which was also an interest of mine. Of course in 1958, calculators had not yet been invented so that logarithms were the only tool available to make the high-accuracy calculations needed for useful trigonometry. At least half, maybe even three-quarters of our class time was spent learning how to accurately use those arcane log tables rather than learning the details of trigonometry.

LOGARITHMS – For those unfamiliar with the term, a common logarithm, aka base-10 log, is the power to which 10 must be raised to obtain the number of interest. For a simple example, the multiplication of 1,000 times 100 could be written as 10^3 times 10^2. The exponents 3 and 2 are the logarithms of 1,000 and 100 respectively, and adding those logarithms 3 and 2 results in 5, meaning 10^5, which is 100,000—the product of 1,000 times 100. Thus adding the logarithms of numbers is equivalent to multiplying those numbers, and conversely subtracting them accomplishes division. Similarly, multiplication and division of logs results in raising to a power or extracting a root respectively. Although it is hard to envision doing the calculation of $10^{3.24945}$, it is still a legitimate mathematical function and would result in the number 1776.

Today, if I were to calculate the height of a tree knowing my distance from it and the angle at which I could see its top, I would use my RPN calculator and just key in the angle in degrees minutes and seconds (38.2100), press a key to convert to decimal degrees (38.35), press the TAN button (.79117), punch in the distance in feet (85.16), press MULTIPLY, and have the answer in feet (67.376066).

Decades ago if accuracy were needed, I would reach for a heavy book (see last paragraph), find the section on tangents and cotangents, decide whether to read from the top or the bottom of the page (38° vs. 51° for example), leaf through 45 pages to the page for the needed degree-value, go down the page to the minute-value listed at left (or right if reading from the bottom), go across the page to the tangent column as noted at the top (or bottom), and read off the logarithm. If there were seconds of arc, there were little charts at the side that were helpful for interpolating between the listed values, but using the chart also required successive computations.

It was necessary to write the 6-digit values on paper for doing the final calculations by hand. Then there was the obscure issue that, for example, the tangent of 38° 21' is 0.79117 meaning that its base-10 logarithm is *negative* 0.10173. Since negative numbers are confusing to include in progressive sequences of adding and subtracting (for multiplying and dividing respectively), the log was actually printed as a positive 9.89827 and it was up to the mathematician to remember to subtract 10 from the final result.

A log table of numbers in the book would then be used to look up the distance to the tree (85.16) to find the logarithm's mantissa (93024), but of course the characteristic (1 in this case) had to be determined by subtracting one from the number of digits before the decimal (2). Thus the logarithm is 1.93024. Then for multiplying the two values, the two logarithms were added (9.89827 + 1.93024 = 11.82851, but remember to subtract 10 to get 1.82851). Lastly the resulting logarithm had to be looked up to find the value for the height of the tree—sometimes called finding the antilogarithm. In this case the antilog of 82847 is 6737 and for 1.82853 is 67.38 for a final interpolated value of about 67.377.

Warning, since these tables were very digit intensive, the typesetters saved space by omitting the first two digits from all but the first column, and left it up to the mathematician to get those two digits from the first column. But in the case of rolling over in midline from, say, 92 998 to 93 003 where only 998 and *003 are printed, there is the need to remind the user to get the first two digits from the next line, not the same line, and this need was marked with that asterisk (as above) or with a bar over the first digit.

My grandfather's 1914 edition of *Logarithmic Tables*, by Baron Von Vega, publishes 8-place logs, thus the respective table listings for the above example are 9.898 2700, 930 2357 for the mantissa, and 1.828 5057 for the final logarithm. This book has 575 pages of data including 270 pages of sine and tangent logs with nearly a million typeset digits, and 100 pages of logarithms for each second of arc over the range of 0 through 4 degrees.

All useless in the face of a 3-dollar app for your smartphone.

There are more arcana, but this is enough for now. And, believe it or not, I actually liked understanding all these obscure details and applying them accurately—after all, I've remembered this much for 60 years. My only problem was an *English* problem: keeping *characteristic* and *mantissa* straight.

Speaking of *English*, my English teacher couldn't believe I had been accepted at MIT and urged me not to go. She undoubtedly thought about my deplorable literary performance and felt it boded poorly for success in engineering school. Fortunately, I did not take her advice, but there were moments over the next three years when I wished that I had.

Eventually, the end of the term arrived and it was time for graduation. It was a bittersweet time: excitement about our new futures but sadness at the prospect of leaving all this behind. Thus I ended my southern childhood.

YOUTHFUL EXPERIENCES, TALES, and ANECDOTES

This chapter, like its companion, *Adult Experiences, Tales, and Anecdotes* found later, is a collection of various issues and stories that had no other obvious place to fit. Some, like learning from my grandfather, the design and construction of an electrical test instrument before adolescence, or the mastery of a surveyor's transit, are germane to my life as a technologist. Others like bookbinding, Scotch tape, and photography are only peripherally connected. While some like Roosevelt's death, making a silver ring, and the praying mantis, are simply ventures from my life that I treasured or had fun with. I hope you enjoy reading or skipping them as you see fit.

The President's Death – On the evening of April 13, 1945, Dad drove us all to the railroad station on West Trade Street in Charlotte to see the train carrying the body of Franklin D. Roosevelt as it returned to Washington from Warm Springs, Georgia. We stood at the southern end of the station where there was a huge brick-paved driveway leading from the street to the tracks. I remember looking down at those bricks, laid on edge and polished with years of freight operations, for they were the only things a five-year-old could see from within a pressing crowd of adults. We were, perhaps, a hundred feet back from the tracks, but Dad hoisted me up on his shoulders for a very clear view as the train approached and passed slowly by the station. I clearly remember the boxcar with its fully open doors, the well-lit interior, the casket, and two servicemen standing at attention, one at each end.

Old Man – My grandfather had a great influence on my ultimate career as a hands-on engineer and has already been described briefly in *The Beginnings* chapter. But I wish to add some isolated and fond memories of things learned from him over my first twenty years.

He could and did make many things. I've described his drill press, screwdriver, and wooden wheels, but there were many other tools or adaptations of tools that I saw him use. He had several stepladders around the house, all of which he made or modified, and all of which had three legs. He said, and accurately so, that having three legs was the only way they would not wobble on an uneven surface. I watched him make a four-footer for us.

Behind his garage he built a contraption for sifting the ashes from the furnace and fireplaces. He put the cold ash on a tray made of hardware cloth, turned a crank that vibrated the entire works so that the fine ash fell through the slightly-inclined cloth while the coarse clinkers fell off the end. The fine ash was put on the garden for fertilizer—potash, *i.e.* potassium.

As I got older, I remember sitting often with Old Man on a swing in the back yard. Of course he had built that swing himself: rectangular steel bars at each end and center to form the shape of seat and back, wooden slats about 2 inches wide and 5 feet long bolted to the bars, similarly fabricated arms at each end, and chains to suspend it from an angle iron A-frame. He made at least two others, one for my Dad who virtually lived in that swing in the sultry Charlotte summers for the later years of his life.

During the hours I spent with him on that swing, he taught me many things. Several times, I watched him use the blade of his pocketknife held square to the skin and drawn slowly back and forth across a finger or other patch of skin to remove a splinter. He told me that the sharp edge of the blade would usually catch on the splinter and pull it out.

He once told me of a bin-like shed he designed for storing cottonseeds. Upon its first use, the shed's walls burst open overnight. He reasoned that the seed mass not only expanded due to the natural heat generation, but also settled during the night. Hoping the two effects would cancel, he had the shed rebuilt exactly the same but with the walls canted slightly inward at the top. The shed served its purpose for many years. His instinct was correct.

Old Man once gave me some advice that I never followed. As he grunted and struggled to get up and out of that swing, he said in a clear voice, "Boy. Don't get old."

My grandfather Hammond is worth writing about since he probably had more positive effect on my life as a practical, inventive engineer than any other single person. Just what would I have been without having known him?

Architecture – When quite young I developed an interest in architecture. Presumably it was combined with or developed from my interest in watching houses being built. The Sunday paper always featured a house rendered with perspectives and a floor plan. I would pore over those floor plans, studying not only the house, but the details of how floor plans were drawn. I designed a few houses myself. When in the fourth grade, Mother took me to visit the architect who had designed the elementary school that had just broken ground nearby. I was already watching it being built and would attend it the next year. The drafting boards, the big rolls of paper, the blueprint machine, and the detailed plans of the school were all fascinating. Among these drawings there was a lot more than just a floor plan layout: there were elevation drawings, details of many features, lists of components, and dimensions galore. He told me that the school would cost $11 per square foot.

The next year I spent several afternoons in the architectural office of a neighbor actually sitting at a drafting board and tracing a church that he had recently designed. One of the nearby draftsmen looked at my work, complimented it, and loaned me a French curve to draw the scrolled lintel over the front door. What an experience.

I continued to plan projects of all kinds on paper, not just buildings. I drew plans for that 7-foot house I built in the back corner of our yard, the electrical test box, the photographic enlarger, and the covered run for the pheasants and guinea hens. In high school, I took three semesters of mechanical drawing and one semester of architectural drawing. The teacher recommended me to an architect who needed a helper for the summer. Sadly for me, someone else got the job.

Scotch Tape – Scotch tape was invented and first marketed in the 1930s but not widely until after World War II when I was still a youngster. I was an early-user and was given a heavy-based dispenser for Christmas when five or six. The tape was glossy and $3/4$ inches wide. I loved to make things using it, and the hands-free dispenser allowed me to make complex assemblies. I sometimes wonder if I would have become an engineer without that convenient access to constructing early inventions and building models. I even made small transparent tubes about $3/16$ inches in diameter and two inches long for holding tiny maps that I liked to draw. The first layer was rolled sticky side out, then it was overlaid with a second layer, and finally trimmed at each end. Those are still among my treasures.

Bookbinding – In the second grade, I noticed that the binding of our spelling book was different from most hardcover books. Over the course of my investigation, I discovered the details of conventional binding: the separate signatures of 32 pages, the long stitches that held the signatures to a loosely woven cloth, the end papers, and the construction of the cover. I decided to make a book myself. I was already saving issues of the Katzenjammer Kids comic books and used each comic book as a single signature. I used cheesecloth for the binding "super" and stitched it without a binding frame. I made the cover boards by laminating thinner pieces of cardboard. The covering cloth was limp unbleached muslin. Eventually, I had created a book. Over the next few years, I bound or repaired several other books.

    In Boy Scouts, I got a bookbinding merit badge, and for that I spent several afternoons at a local custom bindery where the owner taught me all the "proper" ways of doing the things that I had just experimented with previously. The "final exam" included binding the Bear Cub Scout book.

    My later interest in printing was probably connected to the bookbinding interest. My Uncle Buch, the editor of the *Columbia Record*, arranged for me to have several tours of their newspaper production shop and press. I was fascinated by the Linotype machine with its oversize keyboard and huge magazine of type matrices. I observed and figured out how the variable spacers worked to produce a justified line of type; was shown ligatures and where they were found on the keyboard; saw how the typesetter corrected typos; and watched as a line of set matrices was cast into a slug of lead alloy. The typesetter didn't have to proofread backwards: the matrix read right; the molded slug and layout frame was reversed; the impression onto heavy paper read right;

the roll-shaped lead plate molded from the impressed paper was mirrored; thus the newspaper read right. I also was shown the coded notches in each matrix that was used to get it back into the correct slot in the magazine. Years later, I realized that it must have been a binary code, which meant that the Linotype machine could sort up to 127 different characters with just seven notches.

Decades later after my daughter had written so many remarkable emails from her Peace Corps site in Guatemala, I produced one more book, this one from processing and formatting her text, printing it into 24-page signatures, and finally, sewing and binding the book as a hardback complete with colorful headband. And, now, *this* book.

Optics – I found optics to be interesting before turning twelve, and probably the first insight was when I noticed the tile floor in the bathroom appeared to be floating. I figured out that my eyes had crossed slightly and that two adjacent tiles were merging as one, thereby creating an offset in the perception of stereo vision that did not agree with where I knew the floor to be. (I once observed my son discovering this phenomenon, but he was too young at the time to maintain his interest once his eyes flipped back.)

Actually, my first optical experiment was much earlier before turning five when I stuck the lighted end of Dad's flashlight into the bathtub water to see what would happen to the light. I don't remember anything about the light under the water but do remember Dad's fury when the light bulb blew out.

I studied the meniscus lens in my Donald Duck camera when I disassembled and reassembled that. When about twelve after getting my Kodak Pony camera, I developed a good understanding of a camera's optics. The school's Speed Graphic I used in ninth grade was a real joy for me to further examine, and its bulletin included a cross section diagram of its multi-element achromat or apochromat lens, which I didn't understand but found fascinating. I also studied the condenser optics in the darkroom's Omega enlarger.

In tenth grade biology, I did a special project on the eye and its optical behavior. About the same time, I built an enlarger for my darkroom. Years later at Polaroid, I wrote up and distributed an 8-page summary, "General Photographic Data for Engineering" that included nomenclature, lens-image equations, power of a lens, depth of field, depth of focus, preferred apertures, filters, prisms, shutter efficiency, and a graphical chart of ASA film speed vs. brightness (candles/sq ft) vs. EV-number overlaid with relative aperture (f/-stop) vs. shutter speed vs. EV-number.

Early on, I puzzled over the effect of my mother's glasses on my own vision, and I once asked my grandfather why his glasses had an extra lens at the bottom. He told me it was a stronger lens for reading. At some point, I dared to put fingerprints on his glasses just to feel the extra lens and discovered there was no tactile evidence of the transition from the large lens to the inset bifocal. I had reasoned enough about optics to expect a different shape and a step at the straight upper edge. (Some years later I saw a display that explained the manufacture of bifocals using three pieces of glass with different indices of refraction as they went through the many steps of shaping, fusing, grinding, and polishing. All that has been replaced by the one-piece molded plastic lenses of today that, incidentally, *do* have that transitional step.)

Speaking of optics, years later when I was watching a grainy TV channel with Dr. Land (see the Polaroid chapter) he remarked that the "snow" would go away if a small artificial pupil were held in front of the eye. He demonstrated how to make a crude artificial pupil by pinching the thumb and first two fingers together leaving a small triangular hole in their midst. The result was amazing. When asked for an explanation, he paused for a moment then said, "I shouldn't tell you—better to *preserve the wonder.*" I sometimes *wondered* if he didn't know either!

Photography – When eight or nine, I was given a Donald Duck box camera. It had a rigid plastic bellows-shaped snout and used paper-backed film. I loved taking pictures and took many of my friends, our yard, places we visited, and vacations. I also dissected and reassembled it once to see the lens and shutter.

For Christmas in 1951 when nearly 12, I was given a Kodak Pony 35 mm camera. This was true delight. It had adjustable shutter speeds, f/-stops, and focus, all of which I quickly mastered. By the following summer I was sometimes taking color pictures on slide film, but mostly I preferred black and white.

> **NO LIGHT METER?** - In those days, cameras did not have built-in light meters, and I didn't have a separate meter at the time, so I had to learn how to estimate exposure. Exposure for outdoor pictures could be accurately set based on *conditions*. If you set the shutter speed one click up from the ASA number, the apertures were memorized: f/16 for bright beach or snow; f/11 for typical scenes in bright sun; f/8 for those in hazy sun; f/5.6 for bright-cloudy; f/4 for dull-cloudy or open shade. If necessary to change the aperture or speed, the two could be readjusted in tandem to preserve the exposure. For flash pictures, the flashbulb's *Guide Number* was divided by the distance in feet to get the f/-stop while the shutter speed was set at about $1/30$ of a second—just long enough to catch the flash.

In ninth grade I became the school's photographer using a Speed Graphic press camera. I learned to develop sheet film and make enlargements in the school's darkroom.

The following summer I upgraded my own camera, exhaustively researched the options, and purchased an Ansco Regent with faster lens, more shutter speeds, and a bellows to permit folding compactly. That was about the time I began developing my own black and white film and making prints using a print box or enlarger, both of which were of my design and construction. It wasn't long before I began loading my own film cartridges from bulk film.

In graduate school, I bought a Miranda DR camera, an all-mechanical 35 mm single lens reflex. I had long envied the SLR for its ability to view exactly what was to be photographed, its typical acceptance of extension tubes for taking very close pictures accurately, and its easy interchangeability of lenses. Within a year, I purchased a second body so that I could selectively take either black and white or color pictures with a single set of lenses. I also acquired a 300 mm telephoto lens and made the hardware for mounting a 105 mm lens cannibalized from an old sheet-film camera.

Shortly after joining Polaroid, I stepped up to a Calumet 4x5 view camera with its swings and tilts, rises and drops. For several years, I was spending many hours enjoying artistic photography, its travels, technology, chemistry, and finishing. More on this period can be found in the *Outside Block and Polaroid* chapter and the *Yosemite Workshop* section, both below.

I have continued using photography ever since as a means of communication and keeping records, but after my late twenties, it faded as an artistic and passionate hobby.

The Basement – As a youth, I spent a lot of time in the basement. First, it was a play area where Bill and I and friends frequently ran and played when it was too wet or cold to go outside. In my earliest project days, I used Dad's workbench to make things. It had two vises and there were lots of saws, files, and other tools available. Even when I was building structures outdoors, the tools and nails had to be retrieved from the basement. There was another bench that eventually became mine. When I took up raising chickens, the brooder was in the basement; the young caged chicks were kept in the basement each night; the cleaning and dressing of slaughtered chickens was done at the big laundry sink; eggs were stored in its coolness; and I kept the incubator in the basement. When in the fourth grade, I spent about two days watching a new furnace being installed and a slew of new duct pipes run to the registers.

In 1951, Bill initiated the construction of a shower room in the basement on which we briefly worked together. That project is described below.

In about 1955, I adapted that shower room as a sometime-darkroom. I built a cabinet under the sink to store chemicals, trays, and developing tanks. I designed and constructed a contact printer of $^3/_{16}$-inch plywood, a light bulb, glass top, and a rubber-lined pressure-plate lid. When I discovered that 1 by 1.5 inch prints were truly unsatisfactory, I designed and built an enlarger of pipe, wood, opal glass, and tin cans. For a lens, the enlarger held my entire camera with its shutter held open with a locking cable release.

Thus I spent a lot of time in that basement, came to expect and enjoy the smells and dampness typical of them and have found contentment in basements ever since.

Summer Vacation Time – For most youngsters, summer was a time for play. For me, it was a time for projects—planning, building, repairing, improving, *etc*. Even when we went to the beach for a few weeks of vacation, my clearest memories are of venturesome trips to fishing villages, boat yards, construction sites, dredging operations, making sandcastles on the beach, and forming towns and roadways in the damp sand in the shaded crawlspace of our rented house.

When ten years old, I went to Camp Sea Gull for a month and returned there for the next two years. My favorite activities centered around water. Swimming and diving were not only daily, but sometimes multiple times for me. We had swimming and diving lessons and occasionally competitions. But my favorite water activity was boating, probably because knowing how to operate a motor- or sailboat included the challenges of observing and figuring things out. The camp had a fleet of Lightning sailboats. At my age, a counselor always needed to be with us, but I learned all the basics. During my second and third years, the camp had acquired a fleet of Sailfish boats, and a buddy and I took one out almost every afternoon. They also had motorboats which I often enjoyed, but not to the same degree as sailing. My third year, the camp bought a large, perhaps 50-foot yacht, which they used for overnight trips along the Intracoastal Waterway to Beaufort and out into the Atlantic. When they had a naming contest for it, I submitted *Salty*, won, and was awarded with Dixie cups of ice cream for the entire cabin.

The hobby shop was also a favorite, and I built several model cars, the jalopy being the only one that has survived for six decades. Archery and riflery were favorite land activities for me, and I was pretty good at both.

The summer of 1951 was especially busy for me. I was at camp for a month, vacationing on Bogue Sound with family for two weeks, away for nearly a week with my aunt climbing Mt. LeConte, and helping Bill start the shower room project in the basement. But my most significant activity was building my 7-foot house in the back corner of our yard. Despite all these time diversions, I managed to finish the house within the season.

Electrical Test Box – I started repairing electrical appliances when about 10 years old. At some point, I designed and built a testing box to aid me in that activity. I didn't even know about voltmeters.

> **DETAILS** - I designed the box to be made from walnut-veneered plywood. It measured about 12 by 8 by 3 inches high, had a purchased handle for carrying, and included a compartment about 12 by 3 for holding the coiled-up test leads and power cord. The compartment had a removable cover that was held in place along one side with two dowel pins made from brads, and a spring latch at the other edge that I formed from galvanized sheet metal. The box included an otherwise-unused train transformer for low-voltage testing, as well as 120 volt features. The latter included a cord to be plugged into the wall, a duplex socket for plugging in the appliance under test, a lamp socket for either a fuse or light bulb, and two switches. If I needed to test an appliance to see if it had a short circuit, I could install a bulb in the socket, plug the box into the wall outlet, and plug the suspicious appliance into the box's outlet. If the bulb glowed brightly, the appliance was shorted; if it glowed dimly it wasn't. To verify the latter, I would install the fuse in place of the light bulb. There were two ordinary wall switches working together in some function that required a pivoted contraption made from sheet metal to limit them from being in the wrong relative positions.

This box would never have met Underwriter's Laboratory requirements. The test leads were fashioned from heavy-gage galvanized bailing wire, stiffened with wood dowels, and wrapped generously with electrical friction tape to form their handles. The wood of the box served as electrical insulation in a few places.

As with many of my projects, I got far more satisfaction from its design and construction than from its ultimate use. However, I *did* use this one on several occasions.

The Basement Shower – Our house, built in 1932, had only one full bath and that one did not have a shower, only a tub with two walls of wood, plaster, and wallpaper. As my older brother, Bill became a teenager, he wanted a shower and lobbied for something to be done. Dad either didn't want to spend the money or could not envision how it could be done.

In the winter or early spring of 1951, Bill and I joined forces to come up with something. We selected a corner of the basement that had not been fully excavated. The dirt was only about two feet above the basement floor and held in place with a low brick wall. The site was close to a laundry drain. Being a budding architect, I drew up some plans: break a door-wide passage through that brick wall, remove the dirt, use concrete blocks to build two new walls, pave the floor around the drain fitting, and install hot and cold plumbing. The drawings were simple and anticipated that the interior would be unfinished brick and concrete block.

Bill and I began digging out the dirt into a wheelbarrow and hauling it away. The wheelbarrow had to be rolled up a long 2 by 12 plank over the four basement steps. Progress was slow, but we took turns and removed most of the dirt before going to camp.

But, our puny efforts were rewarded because they piqued Dad's interest, and he hired a handyman to complete the project. It was even better than the original crude plans: a toilet, plastered walls, and a single-handle control valve for the shower—a first for our family.

Lathe and Shop Setup – I had a nice little shop setup in one corner of the basement. There was a second workbench in our basement that ended up as mine. It was adjacent to a vast array of shelves, a few of which I used for storing tools and unfinished projects. Our only power tool was a $1/4$-inch drill with drill press bolted near one end of that bench.

When I was 13 or 14, a neighbor gave me an old and badly rusted wood lathe from the 1930s. I bolted the lathe near the front of the bench, and the motor at the back in a manner to loosen the V-belt by lifting the motor. I didn't learn about the trick of flipping the running V-belt over the edge of the pulley with a stick until visiting the Steuben glass factory several years later.

What did I make with that lathe? A lot, including several wooden bowls; a lamp base of laminated redwood; a one-legged stool; all the beads for an abacus; and a barstool with three legs. The stool was about three feet high, so the turned legs complete with beads, hollows, coves, and fillets for ornamentation required the maximum reach of the tailstock. The six rungs were similarly ornamented.

The '48 Ford Deluxe – In 1955, Dad got a new car and turned over the 1948 Ford for Bill to use. His move may have seemed altruistic, but I think he just wanted us teens to leave *his* car alone! Bill didn't get much use of it before he was off to college, but he taught me how to drive and asked me to keep it in running condition for his return. I was still six months away from getting a license but was happy to drive it back and forth in the driveway for half an hour or so every afternoon.

Ultimately, I got my license and used the car for going to school. Bill didn't return to Charlotte in subsequent summers, so it was largely my car for several years. I took good care of that car, performed routine maintenance, and undertook several repairs. I even read the repair manual cover to cover. The starter motor stopped engaging the flywheel, but I probed the problem, replaced a torsion spring, and fixed the starter. Later, the generator stopped charging, and the shop said it was "completely shot." I took the car home, opened up the generator and discovered that one of the brush springs was broken. A new pair cost 15 cents.

I outfitted the car with turn signals...twice. I modified a multicontact switch removed from some Air Force surplus equipment, made a sheet metal enclosure for it that clamped on the steering column with a hose clamp, and ran wires front and back. At the front, I replaced the single-filament bulb and socket in the parking light with a dual-filament, but at the back since there was already a dual-filament bulb, I had to bolt on running lights intended for a truck. One of my friends severely ridiculed the add-on lights, so I removed the running lights and reconfigured the switch to turn off one of the brake lights whenever that filament was needed for the turn signals. I installed pilot lights in the dashboard for displaying the turn lights, but I chose to wire them to the rear light circuit so that the brake lights also lighted up the dash—cool!

The first summer after getting my license, I managed to crash the Ford into the side of a car that ran a red light. The insurance company considered it totaled. I asked for it to be towed to our driveway and undertook its restoration. From a junkyard, I bought a front end for $30—the grille, two fenders, lights, bumper, radiator, and hood. They delivered it. I removed the crushed section and installed the newly-acquired parts using 89 cents worth of fender welting to prevent squeaks in the joint, and had a working car again.[1] Dad painted it black to match the rest of the car. My cost: $30.89.

During winter storage during my first year at college, not only did one of the brake cylinders develop a leak, but so did the master cylinder. (So much for the engineering principle that *two* independent problems do not occur simultaneously.) After a lot of swearing, I bought new seals, brake fluid, and

---

[1] The complete project also required some body work on the two quarter panels. This was before body putty was available, so a neighbor taught me how to finish it off with lead. He loaned me his wooden paddle and body file, but I already had the blowtorch. This was a true blowtorch, not today's propane torch. It used gasoline, and the torch-head needed to be hot to vaporize the gas. This required building a fire under the head for a couple of minutes before using it. Startup went like this: First use the built-in pump to pressurize the gas chamber; then open the valve while using the palm of the other hand to divert the stream of gas into a tray-like cup below the head; light the gas in the cup using a match held with that gasoline-soaked hand; wait for the fire to go out; turn on the torch again; and use another match to light it. No wonder propane torches were invented!

repaired them. It was back on the road, and I used it for the remainder of the summer, but with one additional problem.

Toward the end of the summer, the tires were going seriously bald—so much so that the inner tube could be seen through a small hole. I was averse to spending money on new tires shortly before selling it for only $50, so I patched the tube, placed several layers of cardboard between the tube and the tire, and pumped it back up. That served for about a week...then again...then yet again. I freely told my work crew, including the future buyer about my escapades, mostly in hopes that he would offer to replace the tire even before his purchase, but he just laughed along with the rest of the guys.

Making a Silver Ring from a Quarter – One of my first girlfriends wanted a friendship ring. She pestered me so much that by the time I bought her one, I resented the ring and its one dollar expenditure. For my next girlfriend, I *wanted* to give her a ring although she never asked. I decided against a cheap friendship ring in favor of making the ring myself. In those days, US coins were made of 90% silver—almost sterling. I started with a quarter, used a ball peen hammer to mushroom the edge little by little into a flange of the right diameter, drilled out the center, and filed it smooth and free of burrs. I used a buffing wheel chucked in the drill to polish it. It was beautiful, she loved it.

Slide Rule – My high school never taught us about the slide rule, but when I was in about the tenth grade, Dad showed me my grandfather's slide rule and how to use it. Because of my interest and quick study, he let me keep it at my desk and later take it to college with me.

>**CLOSEUP** – Grampa's slide rule was a relatively simple 10-inch model with A, B, C, D, K, and CI scales. It was manufactured by Dietzgen (Eugene Dietzgen Co.), made of wood (probably mahogany) but faced with ivory into which the divisions were engraved. The slide could be turned over for using the S (sine), T (tangent), and L (linear) scales. The cursor was glass. On the back of the rule, there was a table of 21 useful ratios including square yards to square metres, acres to hectares, US gallons to lbs water, and US gallons to cubic feet. There was also the weight per cubic foot of wrought iron, cast iron, cast steel, copper, brass, and lead. I guess these were the important constants of his day.
>
>When I was about twenty, Dad's sister delivered their father's *Thacher's Calculating Instrument*, a 360-inch long slide rule chopped up into 20 segments mounted around an 18-inch long drum. This slide rule could calculate values with 3.5 to 4.5 significant figures whereas the typical 10-inch slide rule was only good for 2 to 3 figures depending on which end of the scale was being read. I figured out how to use it, but never did so in my work.
>
>I never carried a slide rule dangling from my belt (nor ever wore a pocket protector). Soon after arriving at MIT, I became somewhat protective of Grampa's fine antique slide rule and purchased a similarly simple and relatively inexpensive slide rule from the Coop for daily use. This model from Post (Frederick Post Co.) was made of bamboo with a white plastic facing and used a piece of springy aluminum to join the two halves. That latter feature was ideal for keeping the slide snug at all times. Although inexpensive, it was good quality with engine-ruled markings and a glass cursor. Ultimately, I owned two, one each for home and work. Occasionally, I dreamed of having a K&E (Keuffel and Esser) slide rule with about two dozen scales but generally felt it to be unnecessary. However, in graduate school, I purchased a

5-inch Picket of this style to combine my wishes for an advanced version with those for a pocket sized slide rule. It was especially handy for its Log-Log scale.

But slide rules became obsolete quite abruptly after I had used them intensely for about 15 years.

Remembering pi and e – *Scientific American* sometimes included curious articles about miscellaneous issues. One was about transcendental and irrational numbers, which include π (pi) and the natural logarithm, *e*. The article included a poem of 31 words that aided in remembering π to 30 decimal places. The number of letters in each word represents the value of the corresponding digit. How ironic, the protagonist's name, Archimedes, is too long to be used within the poem:

> Now I, even I would celebrate in rhymes inept
>   the great immortal Syracusan, rivaled nevermore,
> Who by his wondrous lore, untold us before,
>   made the way straight, how to circles mensurate.
> (3.14159 26535 89793 23846 26433 83279)

The magazine also had a poem for remembering the value of *e* to many places, but it was in French, which I did not even try to memorize. But they also parsed it for easy pattern memory to 15 places:

e = 2.7 1828 1828 45 90 45

I have remembered those all these years.

Transit – In June of 1957, my parents and I took a trip northward to tour potential colleges. During that trip, we visited my great aunt in Albany. Her deceased husband, Grampa's brother, had also been a civil engineer. Over the course of our two-day visit, she offered me her husband's transit, and sent me to the basement to find it. I couldn't believe my good fortune as I had often dreamed of using a transit every time I saw a survey crew by the side of the road or on a job site.

> **RESTORATION** – The transit was made by Buff & Buff Instruments, Jamaica Plain Station, Boston, Mass. It had an inverting telescope and a compass that barely turned. The compass dial included its serial number, 4790 and a patent date, Nov. 15, 1900. The transit was housed in its original wooden case with its provenance label pasted inside the door and a few accessories cradled in holders in the case or on the door. I also found the tripod with brass head, oak legs, and rusty steel feet. I didn't find the "chain"—the steel tape measure—until after my aunt's death.
> The transit head badly needed cleaning and lubrication. I found a book at the public library that included a chapter on cleaning and adjustment of transits. I followed that guide to open the two major spindles, clean away the stiffened oil, and apply new oil of the proper type recommended by this book. The adjustments could all be made without reference to outside standards, so I went through each adjustment step whether it was needed or not.

Of course I used it! Over the summer, my buddy John and I ran a survey line from our house to his a mile away. We made stadia measurements of the distance as read with the telescope reticles. I recorded elevation and azimuth angles along with distance from each station. I drew a plan diagram, but never factored in the one or two degree elevation-angle corrections. Over the many years since, I used it as a level several times. Once in the 1980s, I used it to survey and confirm some uncertain markers in our side yard.

Sputnik – On October 4, 1957, The Soviet Union launched the first man-made satellite. At the time, I was a senior in high school, taking physics, recognized at school as one of the "science nerds," and a member of the Astronomy Club in Charlotte. That afternoon, an announcement over the PA system from the principal requested four students, including me, to come to the principal's office after class for an interview. A reporter from *The Charlotte Observer* was there to interview us on the implications of the Soviet's launch of Sputnik. The next day there was a long article along with a picture of the four of us and Mr. Lancaster, the physics teacher beaming proudly at his scholars.

A few evenings later, I attended the Astronomy Club's satellite-tracking project set up at the Mint Museum of Art. The Astronomy Club had joined some Federal program to install tracking stations all over the country for verifying the satellite's orbit whenever one was eventually launched. They had not anticipated an earlier Soviet launch.

> **VIEWING STATION** - The station comprised a tall post, some 30 or 40 feet high, with a crossbar at the top oriented north to south; a series of viewing stations deployed to the east and west of the post; and microphones connected to time-synched tape recorders. Each viewing station had a foot-long section of pipe about 2 inches in diameter welded in place that served as a viewing port from which to look at the section of sky beyond but in line with the transverse crossbar. The pipe was pointed downward for easy peering and had a mirror to fold the line of sight upward to the crossbar. In use, every viewing station was manned by a person who watched for the sun-lit satellite against the early night sky through this round pipe viewer. If the satellite appeared in the viewer field, they were to say "in." When it passed the crossbar, say "pass," and when it disappeared from the field, say "out." Later, all these tapes would be analyzed for exact timing and angular positioning, and the results submitted to NASA. I never knew if NASA was actually without radar capable of tracking and really needed this information, or whether this was just public relations involvement at an exciting moment of space exploration.

Fortunately, I was not a qualified member of the tracking team. Why *fortunately*? I knew that a Russian launch would have been from a site much to the north of Cape Canaveral and would result in a nearly-polar orbit that was not likely to pass through one of our viewing stations intended for a near-equatorial orbit. So I stood to one side, watched the northern sky, and eventually spotted Sputnik heading southeast. I called out, "There it is," and helped others spot it also. I suppose all those official trackers kept their eyes glued to their eyepieces and missed Sputnik altogether.

Michelson Interferometer – During senior year, I wished to enter the Science Fair. I had read an article in *Scientific American* about the workings of a Michelson interferometer. The article included some basic suggestions about how to build one, and of course I loved building things, so I went about designing one in more detail. An industrial hardware store was found for buying precision drill rod, a micrometer head, and fine-thread adjusting screws. I went to a large metal supply warehouse to get $3/8$ inch thick hot-rolled steel, and various sizes of aluminum bar stock for making brackets, legs, holders, and levers. From Edmunds Scientific I ordered glass grinding grit and polishing rouge for grinding the three optical flats needed for the Michelson geometry.

INTRACASIES - Some parts could be made with hand tools at home, but others needed more precision. My high school had no metal working shop, but somehow I solicited the help of a nearby doctor who was also a train enthusiast with a good shop in his garage. He taught me a lot about milling machines and good metal working practices. I went there several evenings and together we made the critical parts. Many afternoons were spent making the optical flats on our kitchen table. I started with three pieces of plate glass about two inches square, successively ground two of them face-to-face with a rough grade of grinding paste, regularly rotating their pairings so they would all end up flat. Then I ground them with successively finer grades of grit and ultimately polished them with rouge using a beeswax foundation for lapping. When testing them against a known optical flat in the physics lab at school, I measured them to be...well...not precisely flat, but hopefully flat enough.

But I also had to put a silver coating on the glass flats, one of which needed to be lightly coated to form a beam-splitter. The *Handbook of Chemistry and Physics* included the recipe for the Brashier process of precipitating silver from a silver nitrate compound. I used that, but went through several cycles to get the process right and to tune up the half-silvered beam-splitter. The chemistry teacher let me demonstrate mirror-making to each of her classes.

Eventually, the interferometer was assembled and working. The not-quite-flat optical surfaces made the fringes somewhat irregular in shape instead of the ideal bulls-eye, but they were entirely adequate for measuring and calculating the wavelength. (One of the judges at the state fair showed a lot of interest, but try as I did, I could not adjust the fringes well enough for him to see. Now that my own eyes have become aged and presbyopic, I suspect that he was wearing reading glasses to see the interferometer clearly, and it never occurred to me to tell him that the fringes themselves appeared to be at infinity. Maybe they were too blurry for him to see while wearing his close-up glasses.)

My source of monochromatic light was the yellow line pair of sodium generated by holding a glass stirring rod in a Bunsen burner flame. At the local and state-level fairs, I used a propane canister of gas and a lab stand to hold the glass rod in place.

After building the instrument, I made a three-panel display entitled "Interference of Light," and included diagrams and text to explain interference, where it can be commonly observed, and how it is useful to scientists. Among my props, I included a peacock feather for its iridescence and a Jews harp because its hardened spring had a beautiful spectrum of colors caused by the interference-thick film of oxide developed during hardening.

At the city-level, I placed well, which qualified me to enter the state fair held at Duke University. Despite the high interest of that one judge at the state-level, I was not a finalist.

It was a fun and rewarding experience.

Praying mantis – In the middle of the summer following my high school senior year, my girlfriend and I were waiting in the crowded lobby of a movie theater when suddenly a girl near us began screaming and pointing at her boyfriend's sock. When I looked, I saw a beautiful praying mantis about four inches long. I could certainly understand her fright and offered to help. But I wanted to keep the mantis, so I wrapped my handkerchief loosely around her and stuffed the gathered corners into my shirt pocket, leaving the mantis-enclosing puff bulging above the pocket while we watched the movie.

The next day I built a screened-in cage about the size and shape of a ten-gallon aquarium. I put a lot of plant material into the cage for the mantis to perch upon and caught various insects that I threw into the cage before snapping the lid closed. The mantis lived there for the rest of the summer. She was fascinating to watch, and among other things I observed that her neck was mobile and she could turn her head to look at me. Although I knew that the mantis has a compound eye, there is something about the optics that gives it the appearance of having a pupil-like dark spot slightly recessed below the surface. Especially in combination with her moving head, it was easy to imagine that she was intelligent and could relate to me through those thoughtful gazes.

SUMMER JOBS
1953 to 1963 – Ages 13 to 23

Not all of my summer jobs were "engineering" or even technical in nature, but they all taught me many lessons including responsibility, working with people, and the pursuit of a good living.

Raising Chickens – From an early age, my brother and I were expected to keep our rooms tidy and help out at mealtimes. Each year more was expected of us. When the chores rose to the level of *contributing* to the household, such as cutting grass and helping to plant or weed the garden, we were paid a quarter an hour. I was cutting grass barefooted with a poorly-guarded mower by age eight or nine and getting paid even though I didn't much like doing so.

But what I did like were the chickens. I helped Mother at the chicken house—picking up eggs from the nests, filling water canisters, adding feed to the bins, and spreading fresh wood shavings. I took over when 13 years old.

For several summers I purchased 100 chicks and raised them until nine weeks of age when they reached about three pounds. Chicks arrive at one or two days old in a carton similar to a pizza box when they still have not had any food. We didn't have a real brooder, but a light bulb hanging in their cage kept them warm. The cage could be carried outside during the day. Although I often slaughtered and cleaned a few chickens at a time, the flocks of 100 each were too large to clean and dress by hand, so I collected them into crates and boxes and took them to a professional chicken-dressing outfit in town. I didn't pay for the feed but I did get paid for each chicken—good deal!

In addition to the 100 fryers, I also kept about 20 pullets all year for their eggs. These were the ones that got slaughtered at home when too old.

As a little boy, one of my play sites was at the confluence of a couple of drain culverts from the driveway and gutters. The eroded and exposed tree roots provided lots of opportunity for roadways for my toy cars, cliff-like house sites, and soon after a rainstorm, a lakeside resort and resource for mud. What more could a little boy want? Eventually, I enlarged that hole until it was a tiny pond about 5 feet across and deep enough to hold water for weeks following a rainstorm. Ideal for ducks. So, one spring I got nine ducklings along with the chicks, and once they were big enough to remove from the brooder, the chicks went to the chicken house and the ducklings to the pond. They were never restrained, but preferred living near the pond. The second year, a dog visited the duck pond and killed eight of them, leaving a single female alive and well, but lonely. I named her Molly Duck, kept her to roam the yard, and collected eggs from her for years. In the spring and early summer she would eat the wild onion tops that often grew in southern lawns, thereby laying me a huge onion-flavored egg most days.

From the time we moved to Park Road, I was fascinated with the guinea fowl (*numididae*) that a neighbor had running loose until they disappeared after a few years. I wanted some guineas too. A doctor in town had a veritable menagerie of poultry including chickens, pheasants, guineas, quail, and bantams in both plain and exotic varieties. He gave me a setting bantam hen along with about 10 guinea eggs—thus the beginnings of my slate guineas with their white-speckled dark gray feathers.

Guineas and pheasants (also from the doctor) can fly so a covered yard was needed. There was an unused, low chicken shelter without side walls that I adapted for the purpose by tacking chicken wire around three sides and down the middle. Attached to it, I built a wire-covered double-pen with each side measuring about 10 by 30 feet.

One of my slate-colored guinea hens. (Photographed by me. Dropping-out the background of the rest of the flock done digitally and recently by the author.)

GUINEA FOWL – The next summer, I tried using an incubator but found it virtually useless for hatching guinea eggs. Back to the bantam. She was named *Banty*, was small, had dark bronze to black coloring, an impressively full tail, and a cocky head. She also knew she was beautiful and was not ashamed of strutting. When she "went settee" on her own eggs, I traded them for guinea eggs, which need 28 days to hatch—seven days longer than her own eggs would have taken. As soon as the guinea keets hatched, I removed them and put 10 more eggs in her nest. She continued to set. Four weeks later, I made the switch again, and she continued to set. After her third clutch of guinea keets hatched, I let her keep them. She was certainly proud of her three-month product, strutted about trying to teach them the ways of the world, but those keets apparently didn't recognize a fancy bantam as their mother for they went darting off in all directions, paying her no mind. She had hatched 22 of the 30 eggs given her—a lot better rate than the one hatchling out of 30 eggs put into the incubator. She repeated this the next year.

I especially liked the guinea fowl. They were fascinating to watch with their helmet-shaped bodies balanced on two sticks, fun to hold if I could catch one, had such ugly necks, heads, horns, and wattles, and made such unusual sounds. Their most common sound was a repeating *ka-trank, ka-trank, ka-trank*, which was the same for both hens and cocks. But whenever they were confronted with something unfamiliar, such as a melon rind newly tossed into their run, they would stretch their necks straight, horizontally, close to the ground, and pointing toward the object, approach it cautiously, all the while making a curious J-sounding buzz.

As I advanced through high school, I lost interest in continuing this business, stopped raising chickens and pheasants but set the guineas free in hopes that although loose and foraging for themselves they might stick around to keep us company. To encourage them I sprinkled feed on the ground daily. They stayed around for over a year roosting in the trees at night.

• • •

From an early age, my attitude toward money was that it was more fun to save than to spend. Mother paid us a quarter an hour to work around the house and yard, paid for the chickens I raised, and gave us an allowance each week to cover school lunches and a bit more. My preference was to see how much of that I could store in a box then transfer to a bank account. I learned about interest, enjoyed seeing the balance grow, and had vague fantasies over what it might someday allow me to enjoy.

Though I did not know it at the time, this attitude would dovetail well in my chosen career. Engineering is a good-paying career, although it does not often make one rich. But it does allow one to consistently put a bit of money aside to take advantage of investment opportunities along the way. And the technical mindset and education gives one the tools to analyze and evaluate such opportunities.

Teaching Swimming at the City Pool: Summer 1955, age 15 – Someone offered me a job teaching beginners to swim at the City pool. My first public service—it didn't pay anything. The program was organized in stations: the first station taught putting the face into water for five seconds and blowing bubbles; the second taught the watermelon float; the third, the dog paddle; *etc.* The children were six to twelve years old and spent as long as they needed at each station before graduating up to the next. We instructors were rotated through the various stations, and I taught at about half of them. No training—we simply worked *with* an older instructor—the apprentice method.

Country Day Camp: Summer 1958, age 18 – The camp was held at the site of a private school. Before the camp session began, the director asked me to plot out all the bus routes. I used a map, push pins, string, and my imagination to plan each route. The buses were 8- or 10-passenger "carry-alls" that had been outfitted with two closely-spaced wood benches for added capacity. No seat belts of course. I ended up driving one—no special license required.

The camp ran from eight to noon five days a week for eight weeks, and I was paid twenty-five bucks a week. I refereed swimming at an off-site lake or worked with the 12-year-olds building a log cabin in the woods. For the swimming sessions—both fun and instructional periods—I would pile a group of campers into a bus, drive it down the road to a muddy lake, supervise for an hour, then return them and take another group. I worked with swimmers of all ages from about 7 to 12. I had a water safety certificate, and other adults were supervising with me.

I don't know who approved cutting down over a dozen pine trees, but that was my assignment as were the details of building the cabin. I quickly learned to use an axe, did not have safety glasses for anyone, and just hoped each tree fell safely *away* from the campers. The cabin was about 8 by 10 feet with a doorway in one 8-foot side. We left the bark on the logs; I chopped them to length and chopped notches for stacking. Some parent volunteers built the roof over a weekend, and it ended up being quite a handsome cabin. At the end of the season, we had a sleep-out using several tents and, of course, the cabin. But there was an unexpected problem: under the bark were thousands of grubs that made so much noise munching their way through the wood that the campers inside the cabin could not get to sleep.

Duke Power Company: Summer 1959, age 19 – The first of two summers, I worked with a crew of four others testing insulators on large transformers and circuit breakers. The group was very friendly, and the foreman was very attentive to safety: he carefully inspected every disconnect knife switch and taped-off the adjacent still-hot units. We traveled in a van (like a stretched SUV) over large parts of North and South Carolina going into outlying substations mostly and occasionally to substations at power plants. (This was the job on which I learned to sleep sitting up in the seat of a moving car with my head drooping.) Our typical week consisted of two 8-hour days and two 12-hour days, one short day in the office writing up the field results. Because of the two extended days, I got paid for 44 hours each week. Another oddity was that one week we would work Thursday through Sunday and the next, Monday through Thursday—eight days on, six days off. What more could a nineteen year-old ask for: a friendly work crew, extra pay, and six-day "weekends"?

> **TESTING** – The transformers were typically 100 kilovolts, about 4 feet square, 7 feet high, and had six large insulators on top angling upward about 3 feet. It was those insulators and the liquid PCB insulation within that we tested. As fresh meat, I was usually sent up top to disconnect the distribution cables and move the test probe from insulator to insulator. I could usually spring up to the top using conduits or other protuberances, but the taller units needed a ladder. The AC test console was inside the truck and was designed to measure the power factor, an indicator of insulation quality. The high-voltage cable with shielded ground was nearly 2 inches in diameter, and the probe was a large sleeve with ground ring and powered hook about two feet long overall. The distribution cables were attached to the terminals with two or three $1/2$- or $5/8$-inch bolts, and the test hook simply hung in one of the bolt holes while a ground cable was clipped from the ring to the case of the transformer. The transformers also had a drain faucet from which we would draw a small amount of the liquid insulation into a concentric test chamber for similarly testing capacitance. The liquid was a phenol: PCB[2] for polychlorinated biphenyl,

---

[2] The PCB story includes a commentary on government regulation. When it was first determined to be an environmental contaminant and possibly carcinogenic, there was such a frenzy of legislation against it that for a while there were mutually-exclusive regulations that a company could not store PCBs on their property and could not transport it away. Companies were in jeopardy of being fined whether they did something or nothing. Every piece of government legislation comes replete with loopholes and opportunities for exploitation and fraud, so a new industry sprung up: "the midnight dumpings." A mystery tanker truck would show up in the night, pump away your PCB, and drive around rural roads for the rest of the night with a drain valve open to liberate the dangerous chemical alongside hundreds of miles of roadways.

later found to be hazardous and carcinogenic. When finished, we just poured it onto the ground.

Circuit breakers were about the same size as transformers and were tested similarly except that the test was repeated for the breaker in both the open and closed position. A few times we tested 230 kV transformers and breakers. On one occasion near the end of a long day, we were called upon to travel to the far end of the territory and wait almost all night to test a breaker that had blown up and was being repaired. While they worked on it, I actually crawled inside the breaker, sat in a pool of that deadly phenol, and peered up at the mechanism, some of which consisted of wood insulators. Got paid well that week.

At the end of the summer, the manager of field maintenance told me that I was a good worker, Duke Power would welcome me back the following summer, and they would assign me to a different challenge.

Duke Power Company: Summer 1960, age 20 – The second summer, I worked with one or two people testing insulation within power generators. This required a 10 kV DC test and effectively measured the charge rate (time constant) of the insulation. The test instrument was a footlocker-sized box with a few gages on top and heavy enough to require two to carry it to the test site. All of our work was in power plants. A few of our tests were at large, modern multimegawatt coal-fired steam plants where the generators were high-speed and horizontally-mounted to an immediately tandem steam turbine. Most of our work was at medium-sized hydroelectric plants where the typical generator was about 8 feet high, 25 feet in diameter, mounted on a terrazzo floor, had about 40 magnetic poles, and was connected via a vertical shaft down to the turbine 30 or 40 feet below. One of the smallest plants was a wooden shack down a steep dirt path from the road. The generator dated to the 1890s, was open-framed, stood about four feet in diameter and two feet long, and was connected to the turbine shaft with a 20-foot long leather belt about 12 inches wide. It was on-line simply because it still worked. It didn't generate much, but as long as it was free energy, why not keep it working? One plant was across a creek on a very wobbly suspension bridge consisting of two cables and an 18 inch wide plank. By walking across, we simulated the *Gallopin' Gertie* failure of the Tacoma Narrows Bridge—scary!

> **SYNCHRONIZING** – It was especially interesting to see how a generator was put back on line. The control room had hundreds of gages, dials, and paper-disk recording instruments, all hooked up by discrete wires or copper-tube pressure lines—no computers and monitors in those days. Each turbine and generator set had a bay of control handles and readouts specific to its operation. For example, to bring a hydro-generator up to speed, the operator slowly turned a wheel about the size of a car's steering wheel to open the turbine vanes, thereby admitting water flow through the turbine. As the generator approached the correct speed, a single-handed clock-like phase meter would display the phase difference between the generator and the power grid. For example, if the turbine and generator were turning at 57 Hz, the phase meter would spin counterclockwise 3 times per second because the grid was at 60 Hz. The operator continued to fine-tune the speed until the hand on the phase meter not only stood still, indicating 60 Hz, but pointed to the top of the dial, indicating that the voltages from the grid and generator were in-phase. At that point, he closed the circuit breaker to connect the generator to the grid. Once connected,

he fully opened the water flow through the turbine so that it could generate its full potential of power. The generator self-controlled the strength of the field magnets to exactly balance the supply-power from the turbine with the load-power being delivered to the power grid. In effect, once connected, the generator remained synchronized to the grid.

The small untended power plants had a curious variation in the above-described method made possible because the generator's capacity was so minute compared with that of the power grid. Whenever we tested at one of those shacks, an operator from a nearby plant would meet us there to unlock the door and operate the machinery. The control instrumentation was minimal, and the phase meter was just a light bulb that glowed brightly if the generator and grid were 180 degrees out of phase and dark if in phase. Such a crude instrument didn't accurately display a zero degree phase angle, and some operators were not even careful about speed synchronization either—when the flicker got fairly slow and the bulb got fairly dim, they would throw the breaker. There would either be a loud clap of noise as the generator abruptly slewed into proper speed and phase or a loud bang as the circuit breaker tripped open. At one of these shacks, the bulb and socket were just dangling from the cables stapled to the rafters.

• • •

A couple of side stories from that summer:

For an abacus-making project at home, I needed some brass rods and asked my foreman if he knew where I could get some brazing rods. The next time we went to a large steam plant he took me to the maintenance shop, introduced me to the foreman and said, "Ask him." I did. He responded, "We can't give 'em away, but if you find some on the floor, you can have 'em." He turned to walk away and as he did so, raked three full length brazing rods off the bench and onto the floor. A useful life-long lesson!

At the Great Falls hydroelectric plant in South Carolina, they were replacing the bearings on one of the hydro-turbines. The access site was a hole in the floor that led about 30 feet down a narrow concrete shaft with iron ladder rungs molded into its wall. I was allowed to climb down the ladder and into the interior of the turbine itself. The control gates did not fully seal, so there was a fair amount of water spilling onto and through the temporary construction floor of planks. The repair was the replacement of the original 1916 lignum vitae bearing blocks that supported the weight of the turbine and shaft.

Raybestos Division: Summer 1961, age 21 – My first real engineering job. After an interview at college, Raybestos invited me to Stratford, Connecticut for an in-depth interview. My future engineering manager told a funny story during our lunch in which someone else was talking about an interviewee he had just had lunch with. The interviewer was asked, "What did you think of the candidate?" He answered, "Well, he was all right...I wish he had shown a little more spark...even if he had just eaten his pie *from the wrong end*, I would hire him in a second." At this point, I was still eating my entree but looked down at my apple pie and wondered if I should eat it as I *usually* did at that time, which was from the "wrong" end. I figured that my usual way of eating would be interpreted as a transparent response to his story, so I ate it point first.

I was hired. It was a very fulfilling summer experience. I had a small drafting board that was nailed to the wall, propped at a slant with 2x2s, and was given real engineering assignments.

The product line of Raybestos was mostly asbestos-based friction products such as brake shoes and clutch plates. My first assignment was a project to figure out why the shaping operation of sheet metal rings was so variable. The ring was about 5 inches OD by 3.5 inches ID by .035 inches thick, and was part of a friction disc used in Chevrolet automatic transmissions. A predecessor had built a press to determine how much overbending was needed to create a permanent waviness to the rings, but he had been foiled in his mission because the results were very unpredictable. It was my assignment to determine why, so I went about checking the press for uniformity, bending a bunch of rings, taking notes about the angular position of the rings in the press, and plotting the results on an 8.5 by 11 sheet of drafting film. They were nice graphs, but the only correlation I could observe was that the depth of wave was consistent relative to the letter **R** stamped on the part.

I had no idea what this meant. But my boss instantly recognized the significance. Problem solved. The R (for Raybestos as the manufacturer) was stamped on the part during the original stamping from the roll of sheet metal stock and, therefore, was consistently oriented relative to the spring metal's "grain." Sheet metal bends differently *across* the grain than *with* the grain and the difference is more pronounced in spring metal. He was delighted with my discovery. Design of the production tool could proceed.

Next they had me draw the plans for constructing the concrete pad for a new 600-ton sheet metal stamping press that was to be delivered in the fall. That took longer because I had to research old drawings of the factory floor, locate the threaded studs to match the hole-pattern of the press, and design a wood template for setting the studs in the concrete. That was probably my first engineering drawing that comprehensively specified an entire job. I did see the excavation in the floor, the pouring of the five-foot deep concrete pad, and the setting of the studs. But the press didn't arrive until fall.

I enjoyed the factory. Many times after lunch I would take my time returning to the office and wander through the factory to watch the various operations. They had some huge machines for forming large steel pieces, a few sophisticated machines, but mostly the factory was served with simple, hand-operated machines for making asbestos mats, stamping or shaping sheet steel, assembling parts, or sanding finished pieces. There were a lot of tote wagons and move-men transferring materials from step to step. Sometimes, I asked my supervisor for explanations of processes I had seen. Other times, I redesigned the machines in my head for improved performance. Once, I drew a diagram of my recent thoughts and shared it with my boss.

After about four weeks at Raybestos, I was told one morning to go see the Chief Engineer in his executive building office at 11 o'clock. I stewed all morning wondering what had gone wrong and what my discipline was likely to be. He greeted me pleasantly, sat me down, and told me how pleased they were with my work. "And the best way to tell you that is to put it in your paycheck. Your salary has been increased from 95 to 100 dollars a week." Wow, here I was struggling, and they felt I was doing a great job.

• • •

As an aside to this story, I had long been aware of the value of money and using it wisely. Throughout those low-paying jobs as a teenager, I always saved some of the money earned and always gave deep thought to whether I should buy a new camera, could afford to buy new wood for a project, or would take a girl to an expensive movie. By the time college rolled around, I had accumulated enough savings that my own money was used to pay for the trivialities of school life beyond the necessities of tuition, room, board, and books for which my father was paying. My high school physics teacher told me that he would double his salary when he resigned from teaching at year-end and became an engineer at the nearby McDonnell Douglas plant. That was not the reason that I chose technology for a career, but it certainly reassured me that I was on the right path for a secure lifestyle. And this incident of getting a five dollar raise to $100 made me proud to have risen above the $5,000 per year average income at that time. And I only had a temporary summer job—pretty good, I thought.

• • •

Over the remainder of the summer, I undertook several more engineering projects. One was the design of a large steel tank for weighing about 100 gallons of liquid at a time. It was to be suspended above head height from the balance beam of a scale. A bolt in each corner attached it to the scale's whiffletree. The project manager suggested, "Oh, well, a $1/4$-20 bolt should be adequate." I thought that was on the small side, but deferred to his judgment. I have wondered ever since how that marginal design survived over the years. Another was for the tooling to support a Boeing 707 water-cooled disc brake assembly during oven-brazing.

During that summer, my only mode of travel was on foot or my three-speed Raleigh bicycle. The bike limited my wanderings, but I did pedal up to the Sikorsky plant, went to the Shakespeare plays several times, and traveled all over Stratford.

I rented a room from a retired couple. Mrs. Broedlin, the landlady had a good sense of humor, and many of the things she said with her thick Austria-Hungary accent were especially funny. We had a lot of laughs that summer. For one example, she told me the sad news that one of her friends had just died. She went on to say, "You neffer know these days who is goink to die next." Pause...then a twinkle in her eye. "Vell, you neffer did know—but it's gettink voise!" followed by her infectious laugh.

She was a kind and generous person. Among other things, every day at 4:30, she drove her husband a mile to the Frog Pond Tavern where he had a boilermaker with his buddies while she just waited in the car.

I had a great summer working at Raybestos, living at the Broedlin's home, getting around on a bike, and dating on foot—my trips to the library eventually focused on Tuesday nights because of a pretty girl working there and with whom I eventually had several dates...on foot.

Hamilton Standard: Summer 1962, age 22 – Hamilton Standard, Windsor Locks, Connecticut, was a division of United Aircraft Corporation and started life as a manufacturer of airplane propellers, but when I went there it had expanded into many other areas, including the design and manufacture of jet engine fuel controls—the equivalent of a car's carburetor. That was my work area for the summer.

My chief assignment for the summer was a design modification for a jet fuel control. The problem was that Boeing 707 airplanes had trouble starting their engines in Mexico City. The cause was the high altitude, the solution was a richer fuel mixture, and the real challenge was to introduce the ability to get that richer mixture without modifying the plane's control hardware or the design of this highly complex, 20-pound, many-thousand dollar instrument. My challenge. I did get some guidance regarding concept, and my design evolved toward modifying a bottom cover plate to include a differential hydraulic piston that would shift about a quarter of an inch if the pilot pushed the throttle a bit beyond the start position. The resulting shift would move a three-dimensional control cam beyond its normal limit to a position that provided more fuel. This design required only a new cam and a newly designed cover plate with piston.

I spent most of the summer on this project, learned a lot about sophisticated design procedures, and found my way through the intricacies of their drawing documentation system. At lunchtime I often wandered around the factory to see some of its operations including early testing of space suits, the manufacture and testing of propeller blades, and lots of highly automated numerical milling machines. My boss was impressed enough with my accomplishments and abilities that as the end of my tenure approached, he tried to talk me into forgetting about graduate school and staying on for a permanent job. When I turned him down, he put in for my replacement. His manager said, "You can't *replace* a summer kid." To which my boss said, "But he's become an effective part of my design team and now I need to replace him!"

For the summer, I shared an apartment in nearby Warehouse Point with another ME from MIT who also worked at Hamilton. I had recently bought a used Zundapp Bella motor scooter that I used for lots of travels that summer including a stop for a 15-cent hamburger at a new, low-budget restaurant called McDonalds—one I had never heard of. I went to Hartford many times, to Stratford once, and through Great Barrington to visit my great aunt in Albany one weekend. I soon figured out how to dress for rain, and could arrive at work almost completely dry despite a hard rain. I did, however, have trouble sustaining a dating schedule with just a scooter.

Scott Paper Company: Summer 1963, age 23 – I was introduced to this job opportunity by my thesis advisor who also consulted for Scott. This site included both the corporate headquarters and a research lab; was next to the

Philadelphia airport; and hired a couple dozen students for the summer. We typically went to the main building for lunch, my haircuts, and a weekly movie *You Are There* or *The 20th Century*, from the TV shows narrated by Walter Cronkite. The personnel department also arranged several sessions for the summer employees to learn about the company in general. The Chairman of the Board even addressed us on one occasion and told us among other things that he raised horses at his farm near the Maryland coast and that his salary was $1 a year. I was impressed.

Scott Paper's primary product line was sanitary paper, otherwise known as facial tissue, toilet paper, and paper towels. All of these were finely creped for the purpose of making them soft. Typically the paper was creped at the end of the first drying stage as the paper was scraped off of a steam-heated, highly polished, cast iron drum called a Yankee dryer. The Yankee dryer was huge: maybe 8 feet in diameter and a little longer than the width of the web of paper, which was about 10 feet.

My project, *restricted channel creping*, was to investigate a novel method of creping paper by pushing it into the throat of a narrow restricted channel. At least two other students were also working on unconventional creping techniques. Our boss was helpful, likable, had a dry sense of humor, and the only time I saw him lose his temper was when one of the other students spent hours key-punching an apparent computer program onto IBM cards that made the chain printer sound like the tune *She'll be Comin' 'Round the Mountain*.

This project was the subject of a patent not owned by Scott that they were merely evaluating. I designed the parts needed to modify an unused machine to push the paper with pinch-rollers into a restricted space, had the parts made, had some uncreped tissue paper made, assembled the machine, and tested it for effectiveness. It worked, but not well, and paper jams were more the rule than the exception. It was an experience, but technically, not my best summer job.

The best parts of working at Scott Paper were the tours of the paper plant, exploring the greater south Philadelphia area, and the social interactions. The latter included having lunch together with the dozens of other college workers from both the research and headquarters buildings.

> **FACTORY** – There were lots of fascinating machines and processes to see in the paper plant about 10 miles down the river, but the paper-making machines were particularly impressive. They were enormous. Sanitary papers are made from fairly high quality wood pulp with lots of fiber content for strength—not the ground-wood found in newspaper. The pulp flows to the machine at about 97% water onto a Fourdrinier wire (screen belt) moving about 10 mph. Suction from below removes most of the water in a few seconds. The wet paper web then transfers via a continuous felt belt to the Yankee dryer where most of the drying occurs in less than a full turn, then is scraped off with a flat edged *doctor* blade, then into a long section of many up-and-down rollers where it is hot-air dried. At the end of the drying section, it free floats onto a pair of huge synchronized rollers where it is cradled and rolled into a multi-ton roll of paper around an undriven cardboard core. There are only one or two operators at each machine.

On one of my visits, I witnessed a paper break. Wow! Seemingly out of the walls there suddenly appeared about six men, each of whom knew his assignment. Half of them were clearing all of the stray paper and pushing it down a hatch in the floor to a stream of water to be recycled. At the head of the machine, a jet of water "cut" the slurry down to about one foot wide so that reestablishment of the proper web path was done with a very narrow strip while the other 9 feet of paper was pushed down the hatch. The other men were maneuvering the narrow web of paper with jets of air from hoses, eventually getting it back in place and onto that cradled roll of paper at the end. Of course, there was a messy section of paper within that roll, but that would be discarded when converting to tissues or rolls in a later process. Once the narrow web was running smoothly, that paper-cutting jet of water was slowly moved back across the slurry to form a fully wide web of paper. It was all fixed within a minute or two. It was impressive to watch such coordination among these skilled and practical craftsmen.

I shared an apartment in Media about ten miles south of Scott and commuted to work through the Swarthmore Woods, a conservation area. The Zundapp scooter was my principle transport and was used for several trips to the outdoor concerts in Philly, exploring the historic district, visiting the University, following the Schuylkill River some miles to the northwest, *etc.* My most fascinating trip was a random one well to the south where I discovered Kennett Square and Longwood Gardens, a former DuPont estate. I was especially enthralled by the variety of indoor plantings there and formed some of my goals in keeping house plants for many years thereafter.

MIT and SIGMA CHI
September 1958 to June 1964 – Ages 18 to 24

When September 1958 rolled around, it was time to head for college. I had packed a huge antique trunk with all my treasures, shipped it off by Railway Express, and headed off to Cambridge with Mother and Dad in the car.

Fraternity Rush Week – During the summer, I had received literature from both MIT and most of the fraternities on campus about fraternity life. Also, I had spent an afternoon learning about different fraternities from a recent graduate of MIT who lived in Charlotte. By the time my reviews were over, I wanted to join a fraternity. The custom at MIT at that time was that rush week, a period of about five days, took place even before freshman classes began. That evening we had our last dinner together, and Mother and Dad delivered me to Burton House where I had a room for a few days. It must have been a bittersweet moment for my parents as they watched me walking away from their life and into a new one. For me, it was filled with both excitement and anxiety, but fortunately I have always been rather laid-back about such transitions and have no memory of any angst.

Saturday morning, September 6, I began working down a list planned over the summer. I visited about a dozen fraternities and never slept in Burton house again. I quickly narrowed my interests to two or three and by the third day was pretty sure I wanted to join Sigma Chi. On Tuesday, I pledged Sigma Chi. Immediately, someone switched on the fire alarm bell, and all the brothers came running to the hallway to welcome me. It was an exciting moment.

Freshman Orientation – Rush officially ended Wednesday night and orientation week ran from Thursday through Sunday. Monday was registration day, and classes began on Tuesday, September 16. Initially, I was registered as being in Course 8: Physics.

Sigma Chi – I moved immediately into the Sigma Chi house on Beacon Street, a block from Harvard Bridge. There were nine of us in our pledge class. We had weekly classes to learn about the history and customs of Sigma Chi, a book to study from, and regular oral tests. There was no humiliating or physical hazing, but as pledges, one of us had to answer the phone promptly, which meant all of us had to run for it, answer the front door bell, and daily vacuum the stairs and hallways. On Saturday mornings, the house manager assigned us a few hours of repair and cleanup projects. We were always expected to wear our Pledge Pins perfectly straight and to be able to answer at a moment's notice any of the minutiae we were studying, which included full name and home town of every active member of the "house"—the fraternity.

FILTHY JOB – Undoubtedly the worst Saturday assignment I ever got was when a sewer pipe in the subbasement sprung a leak. The basement floor in that house was four or five feet below grade at the front, made of wood, and had a crawl space beneath. One weekend, the house manager, a senior, told us that the main sewer pipe from the bathrooms had broken and was flooding into a shallow trench that ran the length of the subbasement. The plumber needed the trench to be emptied. Somehow, the manager knew that there was a drain at the bottom of the trench, but that it was closed off with a threaded pipe plug. We pledges were to remove the pipe plug. Because I was the tallest and skinniest, the plan was that two pledge brothers would hold my ankles, lower me head first through a small trap door in the basement floor, and I would reach down through the sewage, grope around to find the plug, and use a big wrench to remove it. As you might guess, the gloves were not long enough to keep my hands and wrists dry. In fact, my face was pretty close to this muck when I finally found the bottom. Fortunately, it didn't take me long to get the plug loose and watch the sewage recede from my face. I don't remember which pledge brothers I have to thank for their reliability, but surely it was one of many foundations of those lifelong friendships.

The fraternity house had been built as a typical private brownstone residence in the late nineteenth century and acquired by the chapter in 1919. We always assumed that it was built for a wealthy attorney because of the stained glass window between the dining room and the butler's pantry. That window included the image of Shakespeare and a quotation from his *Taming of the Shrew*: "And do as adversaries do in law, Strive mightily, but eat and drink as friends." The building had four floors in addition to the full basement, the first three finished with opulent ornamentation. The simpler finish of the fourth floor suggests that it was originally for storage or for the servants.

The fraternity had two employees, Ethel Franklin, the chef, and Frank Pierce who served meals and maintained order in the house, including making beds after breakfast. Both of them were probably in their 50s. Ethel planned, ordered for, and cooked 19 meals a week. For dinner, we had to wear jacket and tie. On Sundays, there was only one meal: dinner at one o'clock. These two dedicated and respected members of our household worked long hours—about 7:00 to 7:00 six days a week with a few free hours through the day, then a few more hours on Sunday. Frank usually left the house for the afternoon—probably to see his buddies, bookie, and loan shark. For these hours, they were each paid $75 a week.

Undergraduate Years – My courses first semester freshman year were Physics, Chemistry, Calculus, Foundations of Western Civilization, and Astronomy for an elective. Based on my high school experiences with math and physics, I expected those two courses to be moderately easy for me but was in for a rude awakening. Before the end of the first week I knew I was in trouble. The first section of physics was about pucks floating over frictionless surfaces and bumping into one another. Subjectively, this was easy, but the math was beyond me—all those dx/dt expressions at the end of the equations and the expectations that I knew how to solve them. Meanwhile, the math class also had all those dx/dt terms and was of no effective help to me in understanding what to do with them. I was snowed under. Chemistry was coming back to me, but it was being shoveled at such a rate that I was soon overwhelmed by that also. And I never did do well at history and literature, so the humanities course was the usual enigma. My elective, astronomy was the only course that was not sinking me. Every Friday morning, there was a one-hour quiz for the entire freshman class on one of the three technical courses sequentially. So Wednesdays and Thursdays were spent cramming.

For some years prior to my arrival, Sigma Chi had always maintained a first or second position in academics among the 28 fraternities at MIT. They took that record seriously, and I was offered help rather quickly as the chapter learned of my trouble. Peter, the president, took me out to dinner one Sunday night and had a serious talk with me; and his concern and interest helped give me new resolve to master my courses.

I recall a sudden epiphany in calculus class. In the first weeks, I had been completely snowed especially by the instructor's typical use of general notations and hypothetical equations. I was a practical sort of person and needed some specifics to bite into. About six or eight weeks into the term, we got into a section of using derivative equations for optimization problems—one example: calculate the relative height and diameter of a tin can that contains the most food for the least surface area of metal. I was so excited by this utility of calculus that in a single weekend I completely reviewed and learned everything up to that point in the course. Suddenly, I loved math again. I even considered switching out of my physics major into math. My grades suddenly improved, and I was able to do relatively well for the remainder of both terms.

Later in the term, I had a similar awakening when studying integral calculus. In fact, integral calculus became so innate to me that decades later I could usually solve those occasional engineering problems that needed it without having to refer to handbooks.

That fall I also went out for **freshman lightweight crew**. I enjoyed rowing. Despite my typically low athleticism, I have always had good upper body strength, and this sport fit well with that. The resulting blisters weren't any fun, but I got past that and developed good calluses. As the season progressed, the practices got longer and colder. Eventually, I found myself dragging home through the cold and darkness for well over a mile, and into the back door where I could serve myself some cold dinner and eat alone at a stainless bench. Once fed, I dragged myself upstairs to study, which was increasingly difficult given my exhaustion. The first competitive race was on a beautiful sunny Saturday against a heavy wind with waves that frequently breached the gunwales, and we hardly made any progress. We were all "catching crabs" in the whitecaps. As the outdoor season ended but the brutal regimen continued on the machines, I realized that passing my courses was more important than rowing crew and resigned. It was a difficult decision. I had enjoyed it and gotten a great thrill from that first race.

Most brothers had a hand-built bookshelf standing at the back edge of their desk. The few that were available to us freshmen were in tough shape. I designed one to have the features I wanted, somehow acquired the wood, built it of pine 1x10s and $1/8$-inch Masonite, and varnished it. It served me well for the four years I was there and was in excellent condition as I left it behind for someone else.

The closest friends I developed that term were among the upperclassmen. My best friend from our pledge class did not make his grades and had to leave at the end of first term. Two others also didn't make it, leaving six of us to be initiated.

I ended the first term with better than a C-average, hadn't flunked anything, and qualified to be initiated. I could not be smug about my grades, but at least had the solace of not suffering the fate of three pledge brothers.

In **second semester**, my required courses were extensions of those in first semester. This term was even harder than the first. The winter cold while crossing the bridge was unrelenting and we freshmen, although now actives, still had to run for the phone, the doorbell, and vacuum the hallways. The courses were increasingly hard, and the civilization course was impossible. As the term advanced, I felt in just as much danger of failure as in the previous term, but my grades were about the same. When spring arrived, I satisfied my physical education requirement by taking sailing, which I already loved from camp days.

In the early spring, my mother ordered a **preying mantis** egg case to be delivered to me. She did it as a joke, but it was a fun follow-up to the "pet" mantis from the previous summer. It got a lot of laughs, but I rather enjoyed the notoriety. It hatched while I was in class and only one survived my lifeless desktop. I named it Harry, fed it fruit flies supplied from Peter's biology thesis study, and eventually watched him shed his exoskeleton. He quickly grew larger, but sadly escaped soon afterward. Harry was never found.

As the end of my freshman year approached, Earl, the treasurer invited me to succeed him. As then organized, it was a two-year stint: one term as assistant, one full year overlapping summer as treasurer, and one term as comptroller, an advisory position. By that schedule, it always had two overlapping practitioners and left senior year open for higher office. It was a wonderful assignment for me—it recognized my serious, responsible side; it required rigorous adherence to schedule and responsibilities; and it helped develop my self-confidence. I agreed, of course.

One of my most pleasurable memories of springtime at Sigma Chi is of sitting on a fire escape or the roof on a beautiful day. Every room had large windows and direct access to the fire escape, which could be a heavenly place to rest or read. The sun at the front of the house was warm and the river breezes at the rear of the house were exhilarating. The roof was a nice place for sunbathing or sitting on the parapet and gazing over the river and skyline beyond.

Sophomore year began with rush week, this time seen from the inside. During this week I developed a closer friendship with some of my fellow classmen as I observed their sensitivity, caring, and wisdom toward and about the visiting rushees. One of my roommates and I developed a lasting friendship just from walking together across the bridge to class and having 18-minute talks while doing so.

I switched my major to *mechanical engineering*, "Course 2." That first term, I took Physics, Calculus, Applied Mechanics, Engineering Design and Manufacture, and Humanities, two of which were my Course 2 engineering classes. I was still struggling but closing the gap on a *B* average, which made me a lot happier than freshman year.

> **TREASURER** – In my job as Assistant Treasurer, I soon undertook writing most of the checks; publishing monthly house bills for room, board, social, and miscellaneous; collecting payments; making deposits; and keeping handwritten records of every transaction on 11 by 17 green ledger pages. I also had to keep the huge mechanical adding machine on my desk and try to do most adding and reconciling during the day as its noise carried into the nearby study and sleeping rooms. Earl drummed into me the extreme importance of writing and distributing the two paychecks every Friday morning since Frank and Ethel needed to cash those checks on an invariant schedule. Earl continued to compile the monthly summary reports and submissions to our corporate trustees until I became Treasurer one term later. Thus, in my second term sophomore year, I was performing all of the treasury functions. This was before the days of computers and everything had to be done by hand. Although we had custom-printed 4 by 6 forms for the house bills, each one of them had to be filled in by hand after checking for unpaid balances and adding up scribbles from yellow notebook pages for telephone usage, bar bills, and dinner guests. All of these entries had to be replicated on the ledger pages and totaled by hand. Every check and deposit was entered into both the check register and in chronological entries on the ledger pages. And the monthly balance sheet, asset and liability statement, *etc.*, had to be typed onto Ditto masters, taken to an office at MIT to run-off the several needed copies, then mailed out. As the ledger pages were filled up, they were bound into a heavy cloth-covered ledger book.

In the spring, I took Physics, Differential Equations, Applied Mechanics, Thermodynamics, and Humanities. I discovered a struggle with thermo courses, and dropped below a *C* average.

Over the summer, I was living in Charlotte for the last time and working at Duke Power. Accounts needed to be closed over the summer, so I had packed a heavy box with all the accoutrements of the Treasurer's job except for that huge adding machine and shipped it to Charlotte by Greyhound bus. I made do with adding in my head for the summer, but found that each column had to be added about five times before I was sure it was right. I also learned to use an abacus with proficiency and speed that summer, but don't believe it improved overall speed.

Junior year found me taking even more of my type of courses. First term, I took Fluid Mechanics, Thermodynamics, Experimental Engineering, Engineering Materials, Electrical Engineering, and Economics—finishing up with a *B minus* average.

Sometime during this year, I learned that my grades improved if I wore a jacket and tie to class. When I began using the machine shop regularly, I favored bow ties for convenience and safety. Through grad school and for the next fifteen years at Block, Polaroid, and ECA, I always wore a bow tie—self-tied, of course. During my tenure at FMA, I gradually switched to long ties, which I continued to wear until a few years before retirement—"business casual."

Second term I took Dynamics, Mechanical Behavior of Materials, Engineering Design, Experimental Engineering, Machine Tool Fundamentals, and Psychology. Finally, my course work was beginning to truly improve: three *A*s, two *B*s, and a *C* in psychology. The Engineering Design course included an assignment to design a variable speed drive. I really enjoyed this from the selection of a mechanism; the calculation of torques, forces, and unbalanced loads; to the execution of a design complete with shafts, bearings, drive plates, keys, adjustment features, support frame, enclosure, and screws and nuts to hold it all together. Got an *A* for that.

The Experimental Engineering course spanned two terms and was organized for us to form a small group and select a project to complete with the help of an instructor. My fraternity brother Dick and I worked together under direction of Phil, a grad student, to demonstrate an **inverted pendulum—a broom-balancing machine**. Dick's selected specialty was control systems, mine was design, thus we made a good team with me designing the hardware and he the control system. All three of us were compatible and had a lot of fun on that project. We also accomplished a lot. I designed a 3-foot long track, track bed, servomotor mount, drive pulley, and multi-turn potentiometer for position sensing. Also, I designed a cart that rolled in a captured fashion along the track and supported the pendulum, the pendulum pivot, and angular feedback sensor.

Dick and Phil designed the control circuit and hooked it up to a servomotor drive. An MIT shop made the custom parts to my drawings, and I found or ordered all the screws, bearings, low-friction potentiometers, and other standard parts. I also assembled the base, track, pulleys, cart, wheels, and pendulum mount. Dick had the suggestion, "If it works, let's put a glass of water on top of the pendulum bob!" I machined the pendulum bob with that hope in mind—about the size of a hockey puck with a slight depression in the top for the glass. By the end of the term, we had a demonstration device that balanced the pendulum including a glass of water on top.

> **ATYPICAL FEEDBACK** – The control aspects of this project were of special interest. Most automatic control feedback loops have *negative* feedback—when the house becomes too hot, the thermostat "feeds back" a signal to make it cooler; as the amount of water in a toilet tank increases, the float "feeds back" a signal to reduce the water flow. Both of those are examples of *negative* feedback. Imagine either of those examples in which *positive* feedback is used—they would rapidly become unstable: an unbearably hot house or a water-filled bathroom.
>
> But the inverted pendulum included one each of negative and positive feedback. I'll describe it as though it were a broom being balanced in the palm of your hand. As the angle of the broom from vertical increases, the needed feedback is to decrease the angle—as the brush begins to fall to the right, you move your hand such that it is no longer falling to the right. That is negative feedback. However, if you keep this up long enough, you will eventually find yourself approaching the end of the room and need to do something to bring the broom and yourself back to the middle of the room. But if you simply move your hand away from the end of the room—negative feedback—the broom will surely fall toward the wall. The solution is to move your

hand toward the end of the room—positive feedback—so that the broom begins to fall toward the center of the room. At that point, just your action of controlling the broom angle itself will result in your moving back toward the center of the room. Having both positive and negative feedback was one of the reasons this was such a good *demonstrator* system.

However, this wasn't the end of the story. Six months later in the fall of my senior year, I was hired and paid to clean up this project. Although our project was a clear success, the control system part was still in a "breadboard" state. Mismatched cables between the servomotor and amplifier were held together with wire nuts, and the feedback circuitry was built on pegboard in a palimpsest of many modifications, clip leads, and loosely dangling and crudely labeled switches and potentiometers. My job was to make this into a neat assembly that could be used in a classroom by any instructor. I completely rebuilt the circuit board to fit into an aluminum box along with all of the other loose components and a uniform set of batteries serving as isolated power supplies. I found an engraving machine on which I engraved a faceplate for the control box to identify switches and feedback-adjusting knobs. The cables were properly clamped in place and fitted with disconnect plugs.

Both Dick and I got an *A* in that course, and the final dressed-up system looked quite nice and was used for years thereafter. (See frontispiece.)

I returned for **senior year** fresh off of that very fulfilling summer job at Raybestos. I was full of enthusiasm and was mostly taking courses of my selection and expecting to like. Also, I needed to find a thesis topic. By this time, I was **vice president** of the fraternity. Also, I planned to go to graduate school for a master's degree and began surveys and applications.

First term, I took Design Concepts, Hydraulic and Pneumatic Controls, Experimental Techniques, Nuclear Physics, Psychology, and my thesis. I ended up above a *B* average. For the Design Concepts course, we had to design a product to help a handicapped person. There was no list—we had to develop our own ideas, and mine was to design a **pedal operated automobile** for an armless person.

My concept was to provide two shoe-fitting pedals on the floor with which pushing down on one or the other effected turning, rocking the toes forward simultaneously caused acceleration, and rocking the heels downward braked the car. The horn was a big button on the door activated by the side of the knee. Dimming and turn signals were operated by the head. I took the subway to the Dudley Street Station near a Plymouth Fury dealer to make copious measurements and sketches of the car's details.

I developed all the needed features, most of which I designed in some detail. For the steering system, I made a detailed layout of the pedals, their pivot arms, the connecting rods, the custom power steering unit needed, and the connections to the Pittman arm, the carburetor, and the master power brake unit. I also included cross section drawings of the door-mounted horn and the ceiling-mounted adaptor for turn signals and light dimming that the driver could effect by stretching his neck and leaning. Detailed dimensioning of each part was not required. That design got an *A*. I later received the Wunsch Foundation Award for creative design. Also a job offer a few years later.

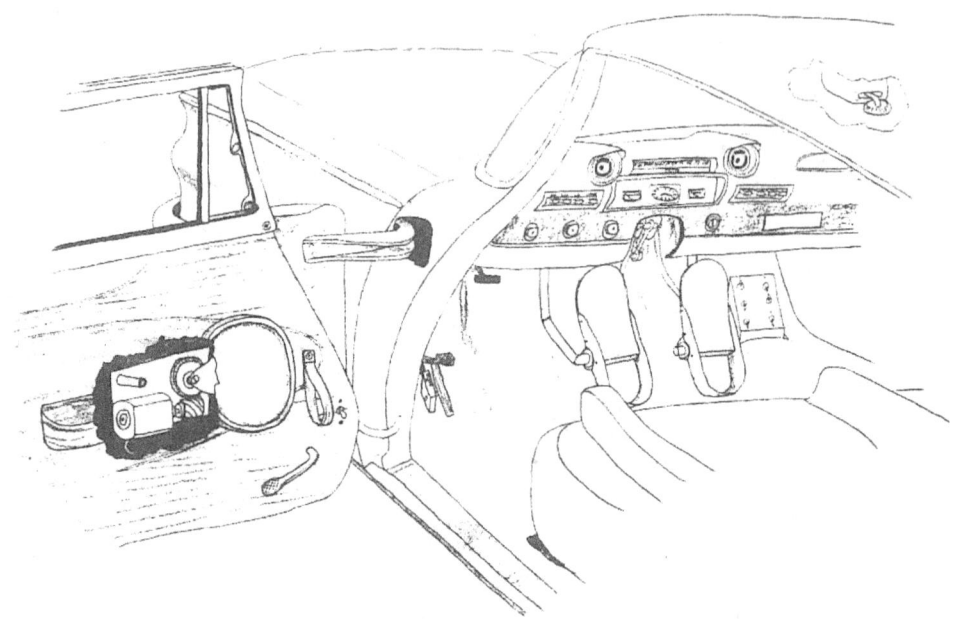

A perspective overview of the Pedal Control System for an automobile. (This sketch was drawn by me in 1962 for a submission to an arc welding contest.)

Bachelor's Thesis – The ME department published a list of suggested thesis topics. Preferring *design*, I focused on the listings under the several design professors. There were many topics of interest, but I eventually selected the design and construction of an **optical accessory for ultrasound neurosurgery** with Prof. Mann as advisor. It had good ingredients for me: design, optics, construction, the neurosurgery connection, and the advisor whom I had already come to like and respect.

    THESIS BUILDING – An early step was to visit Massachusetts General Hospital and Doctors Ballantine and Lele to see what their needs were. They had an ultrasound instrument that focused ultrasound in a converging cone to a point about 3 inches in front of the device. The instrument was mounted to a post in a predictable position and was being used to test tissue-destroying efficacy in cat brains after removing a piece of skull to expose the soft tissue. On occasions, they discovered weeks later that an experiment had been botched because the skull had not been adequately cleared from the field and had caused standing waves and interference patterns in the ultrasound cone of energy. They wanted an optical model of their cone of ultrasound to verify preparedness. That was my undertaking.

    The cone emerged at about 3 inches in diameter and focused to a point about 3 inches away. An f/1 optic would be needed. Their device was quite shallow since the energy source was a piezoelectric crystal only about $1/8$ inch thick. It was going to be a challenge to build an f/1 optical system with so little space. I considered several options, looked into lens sources, drew a lot of rough ray traces, and eventually chose to do it with mirrors. My plan was to start with a low voltage lamp with a small filament, reflect it from a flat mirror at a wide angle to a 4-inch diameter spherical mirror that sharply focused the light to a point about 4 inches in front of it. This plan would not have a solid cone of light, only a peripheral arc, but it was compact enough that it would fit into the same dimensions as their ultrasound hardware. The doctors endorsed the design, but requested a central ray of light and enough rotation to fill in the interrupting shadow from the lamp socket. I agreed and began the design.

After I had completed a composite layout of the entire device, I felt it important to buy or make the spherical mirror early as its availability would make or break the project. I considered several options: Telescope-making did not lend itself to such a deep curve. Edmund's huge stock of mirrors and deeply-dished lenses that could be coated did not include anything close to my need. I visited the local Bosch & Lomb[3] spectacle-grinding shop to see if they could make one or suggest simple methods for doing so. In the end, I chose to make the mirror of aluminum. It could be the ideal size, shape, radius, and include threaded mounting and adjustment holes on the back side. By this time, I had developed enough of a reputation as a competent machinist that the training shop permitted me to work there. That shop also had a radius-cutting adaptor for a lathe. I made the highest quality surface possible using a carefully-sharpened tool followed by successive grades of emery paper for a near polish. In the end, I dared to develop a final high polish with an ordinary buffing wheel gently applied to avoid an orange peal finish. It worked, and I had a bright chrome plate applied to preserve it.

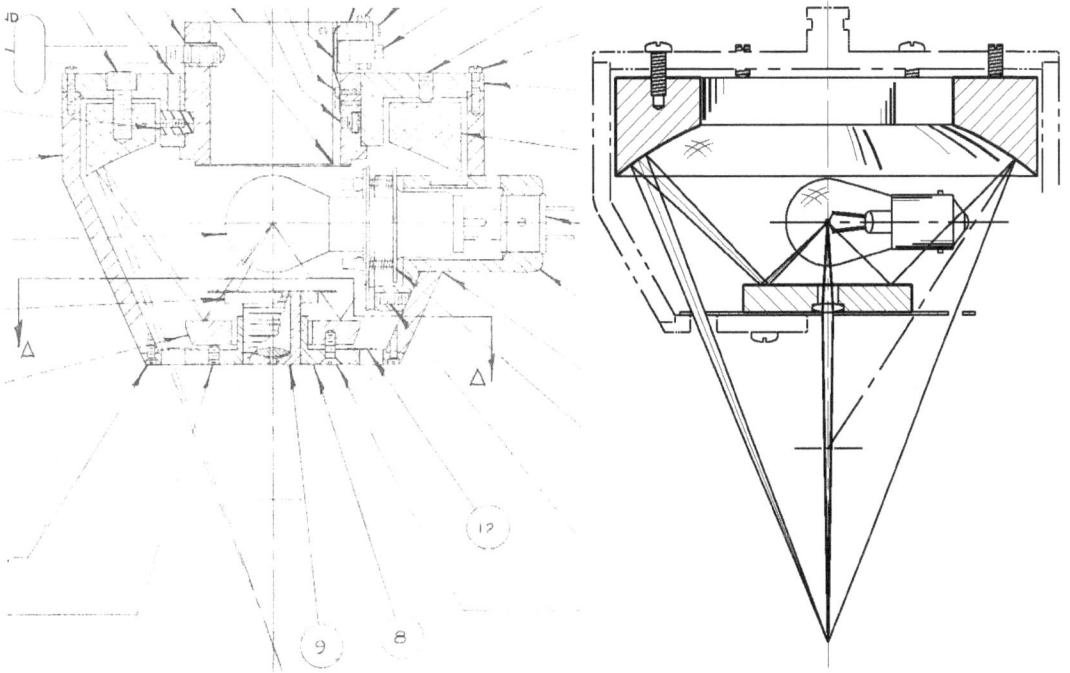

Cross section views of my BS thesis. Drawing on left is copied at 50% scale from an old ozalid print of the assembly drawing done in 1962. The illustration on the right was drawn from memory in 2009 using Cadkey in which the outer case was deliberately simplified. (The similarity of these two drawings should settle any question of whether my memory is still good 47 years later!)

---

3 At the Bosch & Lomb shop I was fascinated with the various techniques of making custom spectacle lenses with prism and astigmatism in a quick and efficient manner. The fundamental secret of efficient custom lens-making was to attach a handle to the already-finished front surface in a manner that determined the prism and indexing of any needed astigmatism. The handle was formed by molding Wood's metal onto the polished surface with a rubber mold. It stayed with the lens through the various steps from rough grinding through fine polishing and assured the proper indexing at each step. When the rear surface was completely polished, the Wood's metal was melted away simply by immersing the lens in hot water.

Once I had made both the curved mirror and a much easier flat stainless mirror and found the ideal prefocused light bulb and socket, I built a little optical bench to test the scheme. When I showed it to Prof. Mann, he was very pleased with the result. The next day he asked if he could borrow it for a few hours. Later that day he returned it to me with a broad smile and announced, "Thank you. This just got you accepted into graduate school." Up to that point, I had been hopeful, but feared that my overall grade point average was not high enough for MIT.

With that challenge behind me, I began to detail the individual parts. I purchased a huge piece of brass through MIT's professional shop—it was 6 inches in diameter by 8 or 10 inches long. Much of that piece was turned into chips as I patiently carved away most of it until I had a shell $3/16$ inches thick, about 3 inches long, and with both cylindrical and conical regions to form the major exterior wall of the new optical instrument. (I never had a *name* for this device and just awkwardly called it "an optical adjunct to neurosurgery equipment.") That large piece of stock was also used for the front face, the rear mounting plate, and several smaller parts. The odd types and various sizes of screws were found at an industrial hardware supply in downtown Boston.

I thoroughly enjoyed working in the shops, making parts, using various machines and techniques, and learning their eccentricities. Eventually, I finished all the parts and assembled the unit to verify fit and function. Once again, I visited Mass General to show it to my clients. They were pleased with the result, so I disassembled the unit, vapor-blasted the exterior surfaces to give it a dull finish and remove the machining marks, and had it chrome plated.

In the later stages of the thesis project, I began to write it up, more or less following the thread of its beginning stages, its evolution through the approach and design options, selection of materials, and its final functionality. It also included parts lists and blueprints of all those drawings. (Actually, the true blueprints with white lines on blue background were expensive to have made, so they only went into the original submission for the Engineering Library as required for archival stability. Other copies, including mine, had ozalid prints, the blue-line ammonia-developed prints that I could make myself but were expected to fade over the years.)

In the second term senior year, I took Engineering Projects, Automatic Control, Electronic Circuit Design, Psychology, and Thesis. Got an A in everything except for that darned psych course.

I hardly remember **graduation**. It was in Briggs Field House, an enormous space. There was the cap and gown, my parents in the audience, diplomas awarded, and a luncheon afterward in the Great Court.

That was the spring when I bought a used Zundapp motor scooter. Soon after graduation, I drove that machine down to Windsor Locks, Connecticut to begin work at Hamilton Standard.

• • •

In retrospect, **Sigma Chi was a lifesaver** for me. I think had I been living alone or with only one other person in a large indifferent dormitory, I would have become overwhelmed with schoolwork and loneliness. Whether I would have given up, flunked out, or struggled to succeed will never be known. But those close friendships developed in Sigma Chi were certainly fulfilling and memorable, and I shall be forever grateful. Moreover, the education and responsibility that I gained while being the treasurer and vice president as well as generally taking good care of my brothers, our employees, and the house itself were good lessons in life for the moment and for the decades that followed.[4]

Graduate School – This period was a mixed bag of experiences. The first year, two other grad students and I rented an apartment in Boston. The food technology major did most of the cooking while the other and I did the cleanup. We got along well, were friendly with the girls upstairs, and used my motor scooter for a lot of traveling in and out of the Boston area. The second year, I rented a room closer to campus, and joined the MIT Choral Society.

I was a research assistant, which meant fewer classes at a time, two years for the master's degree, and getting paid for the very research I was doing for my thesis. My thesis topic was to design and build a Braille reading machine that could immediately convert a punched computer tape to Braille long enough to read one line at a time.

The first year, I was in an office with a friendly grad student working on his ScD thesis. An MIT-employed illustrator was in the adjacent office. He had a good sense of humor, enjoyed photography, and was quite friendly. As I was also getting back into photography with my newly-purchased single-lens reflex, we had a lot to share and made some prints in his darkroom. The second year, my office was moved into a bullpen with many 3-person cubicles. It was much less satisfying and lonely until I discovered a friend with whom I could talk for hours about nothing.

Over those two years I took courses in Strength of Materials, Theory of Material Interfaces, Processing of Materials, Advanced Mechanics, Advanced Calculus, Computer-Aided Design, and my Thesis. My summer at Scott Paper had been much less stimulating than previous summer jobs, so I returned for my second year in a somewhat listless state. Also, a couple of these courses were quite difficult, and probably for a combination of reasons I was losing motivation. Despite the slump, I hung-in enough to get As and Bs. I definitely didn't want to stick around for a doctoral degree.

Master's Thesis – My thesis was the design, construction, and testing of a **Braille reading machine**. The objective was to use punched paper tapes from computers as the source of data for temporarily setting up a line of Braille print for a blind reader to use. This was long before data could be moved via Internet, and my best source for punched tape was a summary of the day's news from the Wall Street Journal. The tapes were distributed by Teletype to local daily newspapers so that a tape-reading Linotype machine could set the type for the local newspaper without having to manually retype it. It was my job to come up with a simple way to utilize such data.

BRAILLE READER – The plan was to start with a standard ASCII-coded punched tape, use a mainframe computer to read it, and punch a new tape in which

---

[4] Shortly after graduation, I learned from the Director of Admissions (who I also knew as our fraternity advisor) that I had been one of those high-risk admissions to MIT for whom they did not have great hopes. I had surprised them—graduated on time, never failed a course, good grades the last two years, and already accepted into graduate school.

the punched holes corresponded to the Braille raised dot pattern. (For reference, the Braille cell is two raised dots wide, three high, with dots spaced 0.090 inches apart and raised about 0.015 inches high.) I quickly demonstrated that a Digital Equipment Corp. PDP-1 computer could do this. For example, the lower case letter "*k*" has an ASCII-value of 107, while the Braille letter *k* has two raised dots at the upper and lower left corners of a vertical 2 by 3 array of potential dots. The final plan of the machine would determine which of the standard holes in a punched tape would correspond to those upper and lower left sites, but jumping ahead, let's assume it to be the first and third hole down from the top of the tape. That combination of holes would have the binary-value of 1 plus 4, or 5. Thus the PDP-1 was programmed to read "107" from the original tape and punch a "binary 5" into the new tape. If the computer read "75," which means upper case "*K*", the PDP-1 would first punch the symbol for upper case at one letter-position, then punch "binary 5" into the tape for the next letter-position to mean *K* because the upper case letters were designated by a preceding symbol, not a different dot pattern.

I considered several means for forming the Braille dot pattern. One was to electrically energize three solenoids, each of which would push a bistable pin into an upward position in some kind of matrix. If the pin were about $1/16$ inch in diameter, had a rounded end, and was pushed up about $1/64$ of an inch, it would feel like a Braille dot. A blind person needs to have a line of letters to read, not just one letter at a time, therefore, such a system would need to have the threesome of solenoids march along the line to set-up pins or have the line march past the threesome of solenoids. Alternatively, a fixed matrix of, say 180 solenoids could set up a fixed line of 30 characters: about five words at a time. None of these alternatives were envisioned to be the basis of an inexpensive Braille reader for the disabled.

The approach I developed was to use the punched paper tape as a mask for dropping from one to six tiny bearing balls into a belt devised to cradle the balls and allow a segment of the ball to protrude through the belt by about 0.015 inches. This had the advantages of being able to create an exact sized Braille pattern and using only one position on the punched tape for each letter.

The outline plan for the reader machine was to have an endless belt that ran for about ten inches across the top for reading, around a pulley at each end, and return across the base about four inches below. The ten inches would accommodate 40 letters, or about seven words. Between the upper and lower stretches of belt would be a mechanical tape reader that took tiny balls from a hopper, drop only one through each hole in the punched tape, rearrange the six-in-a-row pattern of balls into the two by three Braille matrix, and deposit them into the back side of the belt.

CONSTRUCTION – The ball selected was $1/16$ inch diameter; the belt was a composite of very thin stainless steel and a cork layer to cradle the ball. A .0595 inch diameter hole through the stainless allowed the ball to protrude by 0.015 inches, and the corresponding hole through the cork was oversize enough that the ball readily dropped into and out of the recess. A set of gears synchronized the advancement of the belt with the punched tape. The balls were assured to drop only one at a time by having a specially-designed spool between the supply hopper of balls and the paper tape. The circumference of the spool had a continuous circuit of pockets in line with each of the six rows of holes in the paper tape. Each pocket was deep enough for a single ball.

For use, the lead end of the Braille-prepared punched tape was inserted in the tape reader, and a crank on one of the pulleys was used to advance the machine. As the punched tape passed through the tape reader and for each row of holes, a single ball would drop out of the spool, through each punched hole in the tape, and fall down a chute into the belt. The six chutes were angled so that the single row of six possible balls falling through the tape ended up as two rows of three balls when it reached the belt. Of course, a hole-position that was not punched would not release a ball. As the crank continued to turn, the ball-loaded section of belt progressed around the right hand pulley and across the top of the reading machine from right to left, so that the blind person could read the Braille as it passed under his or her fingers. A back-up surface prevented the balls from falling out of the belt while going around the pulley or across the top. With a stretch of ten inches of Braille exposed at any instant, the user could pace their reading even if it did not precisely match the belt speed. As the Braille reached the far left of the machine, the balls would drop out of the belt, through an opening in the back-up plate, and into the supply hopper to be used again.

For this thesis project, I hired an undergraduate student to fabricate most of the parts in the machine shop. I made at least two of the critical parts myself: the tape-reading spool that supplied a single ball to each hole of the punched tape; and the endless belt that cradled the balls and presented them to the blind reader. The belt required about 1,000 holes to be punched—I made a tool and fixture to punch about 36 holes at a time. There were about 40 detailed drawings and several assembly drawings, all of which were drawn, dimensioned, and checked by me.

Did it work? Yes. Four different subjects were asked to use the reader, then interviewed about the experience. All of them found the Braille quite readable, but they also noted that the feel of the smooth metal belt and the raised dots was not the same as that for paper and that its friction made it more difficult to rub and move the fingers forward. As expected, the temporary cranking method of advancement made use very awkward, so for most of the testing periods I did the cranking for them. Dependence on gravity to both set up the balls and clear them from the belt did not always work reliably.

I wrote up the project and delivered a full set of drawings as the work product for my thesis. It got an *A*. The thesis text included a discussion of improvements and additional work needed. The latter included some method of driving it with a user-controllable motor. This thesis was not nearly as satisfying as the previous one—the optical instrument.

A funny aside: By this time at MIT, I had become quite friendly with the manager of the student shop. He was in his 60s, very knowledgeable and helpful about making things, very patient with us students, and a delight to talk with. One day with tongue in cheek, I pretended to puff up and boast that the machine I was building had "over a thousand moving parts," referring to that hopper filled with tiny bearing balls. He looked at me with a stern face, put one hand on my shoulder, and slowly said, "Son…that is nothing to be proud of."

• • •

During my second year in grad school, I was asked by the officemate of the previous year, to lend assistance to an optical problem he had with his project. The assistance included driving out to Diffraction Limited on a Friday afternoon to discuss the problem with one of their engineers. During the drive out, the car's radio program was suddenly interrupted with the announcement that **President Kennedy had been shot** in Dallas. We could not believe it. The meeting could barely take place, and no one was able to concentrate on the issue at hand. We eventually and glumly returned to the office. It was a difficult weekend followed by the shooting of Oswald on Sunday morning. I remember going out driving on my motor scooter later that Friday evening with my girlfriend and how we found the streets virtually empty. It was a sad and unsettling time.

As spring of my second and last year approached, I decided to **see my country** before starting my first job. As I narrowed my job choices, I made it clear that I didn't want to start work until September. As the time approached, I began looking for a car and decided that a Volkswagen beetle would be best: economical, adequate, a good choice for later use in crowded Cambridge, and common enough that it had a good service network throughout the country. I bought a used one for $1,200: a light blue 1961 model with sunroof, three years old with 15,000 miles. I made a giant hook to fit over the edge of the sunroof opening in case the trees were too far apart when putting up my jungle hammock. It was needed only once, and that was in the Louisiana bayous where the trees were too close together.

The trip covered many of the classic places to see in the USA, and I met many delightful people, some of whom I kept in touch with for many years. It turned out to be a truly memorable three months that I have remembered fondly and frequently over the half century since.

BLOCK ENGINEERING
September 1964 to December 1964

During the spring of my second year of graduate school I signed up to meet with a number of visiting recruiters at the MIT employment office. I preferred to work for a small company and narrowed my choices to Polaroid and Block Engineering, both located in Cambridge.

The interviewer for Block was a young, self-confident engineering manager with a sense of humor and broad interests. We bonded, and he was impressed with my high school Michelson interferometer project—in fact, his current project was an **infrared interferometer-spectrometer** of the Michelson design. I accepted their offer under the condition that I could start in September instead of June. I spent that summer traveling the US and returned for work in September.

The infrared spectrometer was to be used during one of the **Gemini space flight** missions to measure 8 to 13 micron radiation from a gas cloud. I wondered if it were actually a probe of Russian territory in search of some nefarious activity. The instrument itself was about the size of a beer can and had a lens of germanium, a metal that is transparent in that IR range but with an index of refraction near 4.0, which made the lens appear much too weak to serve its purpose. That range of IR is emitted by objects at room temperature; therefore the instrument had to be very cold to be useful. The spectrometer was surrounded by a double jacketed stainless steel Dewar with liquid neon at 27°K, surrounded by liquid nitrogen at 77°K. That arrangement preserved the expensive neon much longer than otherwise.

Unfortunately for me, the project had been entirely designed and built before my arrival. I ended up designing a test adaptor, topping-up the liquids each morning, and setting up tests. Despite my requests for more to do, nothing of substance was proffered. I was also beginning to see Block as an unrefined outfit with an unpredictable president, too few projects for stability, and totally dependent on government contracts.

After nearly three months and despite my concern with jumping jobs so quickly, I was in such agony being idle most of the time and at best doing menial work, that I contacted Polaroid, which had offered me an engineering job six months earlier. "Yes, Polaroid is still interested in you, come in for another visit." Which I did, and was again offered a job.

Block ended up as a mere blip in my life—eventually, I even dropped it from my resume. (A recent version of my resume and the later-developed CV are found in *Appendix B*.) The spectrometer did eventually orbit.

OUTSIDE BLOCK and POLAROID
Hayes Street, Cambridge, Massachusetts
September 1964 to September 1968 – Ages 24 to 28

Even when I left work at the end of a day and headed home, I usually remained active and involved in technical or semitechnical endeavors. My interests and activities were often directed toward inventing, creating, designing, making, fixing, building, or improving something in my surrounds. And this was true in the very first home of my own.

When I returned from my summer trip around the country, the first thing I did was find a place to live. My choice was in an odd compound of about seven buildings within a board fence and owned by a 75 year-old woman and her husband, both of whom were lovable, eccentric characters. According to legend, the compound had been established by Fr. Feeney, a Jesuit priest who had been excommunicated in 1953 for too-conservative beliefs. He and his flock had since moved to larger quarters west of Harvard, Mass. The compound was equidistance from the Charles River, Harvard Square, and Central Square. Typically, my address would have been 23 Hayes St., Apt 5, but the system had been set up to write it as 23-E Hayes St. Although I liked the simple address format, it did have one problem—a UPS package was delayed several days while they looked for *East* Hayes St.

The rear entrance to my unit had a covered entryway large enough for several bicycles and two motor scooters, including mine. My door was clearly marked with a label from a Bellows bourbon bottle. The apartment had a big living room about 12 by 14 feet, a 10 by 12 foot kitchen, a full bath with closet, and a small bedroom about 6 by 10 feet.

I soon met the two immediate neighbors. The man below me had recently graduated from Harvard, was very liberal, had no interests in common with mine, and while we were entirely friendly, we never did anything together. Gretchen from upstairs was an attractive Harvard grad student who drove the other motor scooter. We often stopped to chat, sometimes about stock tips, periodically got together in either my or her apartment for drinks, but never dated. I met a nearby family with children for a curious reason: I needed to photograph a very young, blue-eyed child for my Polaroid research on red-eye, and their two-year old was a perfect subject. I have a startling picture of him with one brilliant red eye and one yellow eye.[5]

The **apartment** soon became an eclectic place that was furnished mostly in trash. The weekend after moving into the apartment, my brother Bill

---

[5] The commonly-noticed red-eye in photos is a reflection from the rhodopsin in the retina. The atypical yellow was achieved by having the boy look to one side to precisely line up one optic nerve (blind spot) with the camera. No rhodopsin there, hence a different color.

drove me in his station wagon to an estate sale in Gloucester where I bought a well-worn rug for $5, a poorly-painted bureau for $5, and a whole lot of mismatched table and cooking ware for $3. A bunk-type bed frame with casters was found in the basement. I replaced the casters with 2x2 wood inserts sized to raise the bed slightly and to level it on the bedroom's sloping floor. Over several trash days, I found a sofa missing its bottom cushions (upgraded a few months later with a two-piece sofa *with* cushions but a burned-out backside); a heavily paint-spattered chrome chair; three brown-stained pine boards that with 6-packs of empty beer cans for end-supports formed a bookcase; and two damaged mattresses that, for $35, Goodwill picked up and transformed into a single, extra-length mattress with new cover and sterilized stuffing. A friend gave me an enameled-top kitchen table with drawer. Living room drapes in bland polyester were bought at Sears.

Over the next several months, I made two pieces of furniture. For a coffee table, I bought a single 14-foot 1x10 clear pine board, which was nearly used up to construct a center-slotted top, legs, and leg braces. To make a bookcase, I bought three 10-foot 1x10 knotty pine boards that were similarly used up for its construction. Only tiny scraps were left over. Both of these were finished with walnut stain and satin polyurethane. The AM/FM radio from junior high school was my sole source of music and news for over a year.

Tools? I already had an electric drill bought the previous spring for installing seatbelts in the VW. There was the toolbox from the MIT thesis-building period and the Zundapp engine-rebuilding project on Gloucester Street. I purchased a 12-pitch handsaw from Sears for making reasonably smooth cuts in the above-mentioned boards. The collection slowly grew as needed and some tools were stored in display-fashion on the bookcases.

Almost immediately, I began to **collect plants**. Most were bought at Stop & Shop, larger ones from Lexington Gardens. I had several varieties of wandering Jew and chlorophytum as hanging plants. A dracaena grew to the ceiling, and someone gave me a huge bird of paradise. The bedroom window sill was widened with a tacked on board and littered with philodendron, ivy, and other climbing plants so there was no need for curtains. My philosophy was to water them the way I found easy—if they didn't survive, then I got something else. Eventually, my entire collection could tolerate my watering habits.

To some extent, I also decorated the apartment with trash. Several items, such as a pole lamp, a colored glass fishing net float, and a clear glass retort literally came from the sidewalk on trash day. My ice ax from Rainier was on one wall. The tumbleweed, which had survived the drive from the Arizona desert cradled beneath the back seat, was suspended with white thread glued to the white ceiling with a dab of white glue—it appeared to be floating. I created a sculpture of randomly placed tin cans soldered together. These were mostly cans from meals or juice, a drippy paint can—the handle being the only moving part in the sculpture—and Black Label beer cans, which had solderable steel end caps in those days. When finished, the sculpture was about 20 inches in diameter and 78 inches high, a little taller than I. Over the summer I had acquired a silkscreen print with added brush strokes, which I framed myself. The ugly pole lamp was soon replaced with one of those concertina-like paper lamps and two multiuse photoflood reflectors on tripod stands—one aimed at the ceiling for ambiance and one for a reading lamp. From junk and antique shops, I began a collection of unusual shot glasses, which graced one of my bookcases.

The tin can sculpture I made in the 1965 to '66 period. (Photographed by me using the Calumet 4x5 view camera.)

• • •

Since I had taken lots of black and white **photographs** over the summer, one of my first missions was to get some of them printed. The bathroom was dark enough at night for loading film into a developing tank, and I developed it in the kitchen. I was still a member of the MIT Hobby Shop, which had a darkroom, so I enlarged many of the prints there. I also used the table saw at the hobby shop to shape picture-framing stock from some lauan boards I purchased at a local lumberyard. From Sears, I purchased framing clamps, one of which included a rather poor but adequate mitering guide. I made about a dozen frames while crouched on the $5 rug. Eventually, this kind of rug use dictated that I get a vacuum cleaner, which I did on trash day—it was one of those wonderful 1930s vintage Hoover machines with light, beater bar, and cloth bag. I dry-mounted the prints, inserted them into the frames without glass, and hung them to supplement the plants and trash as artwork.

Not long after moving into Hayes Street, I became serious about **artistic photography** and moved up to a 4x5 view camera complete with a black cloth over my head.

LARGE FORMAT PHOTOGRAPHY – I bought a new Calumet view camera with swings, tilts, rises and drops, a case, and a robust tripod for it. I needed a more extensive darkroom, too. I ripped a sheet of plywood and constructed a 30-inch bench along one wall of that good-sized kitchen, using the 18-inch leftover for a shelf below the bench. The MIT Hobby Shop had a retired 4x5 enlarger that I offered to buy for a fair but low price. It was falling apart, and the only way I could get it to function was to attach the top of the column to the wall and to cut shims from emptied beer cans to square-up various joints. It had no lens, so I used the camera lens as an enlarging lens. I made the safe light from a junk lamp head. A used printing easel and enlarger timer were found at Crimson Camera where two or three used lenses for the camera were also bought. Chemicals were mixed and stored in gallon jugs left over from Stop & Shop wine. The camera, developing trays, paper cutter, and a print washer were about the only items I bought new.

For about two years, I really got into photography. Almost every weekend, I would go out in my VW beetle with the Calumet case on the back seat and look for picture opportunities. Even when going out of state to visit people, I would take the camera. The following week, I would develop the negatives one night, then make prints of the best ones on other nights. I usually used Kodak Plus-X film and D-76 developer, though occasionally used infrared film. The 4x5 film was developed by hand, shuffling about six sheets at a time. I had a chart of developing times at various temperatures and accepted the prevailing room temperature instead of trying to regulate them all to something other than the room temperature. Many pictures were made using Wratten filters, either 25-A (red) or 15-G (yellow). These would enhance contrast, especially of clouds against the blue sky. After I joined Polaroid I was able to use a lot of Type 55-P/N film, a 4x5 film that resulted in both a print and a negative from which I made enlargements.

Not long after joining Polaroid, I sought out Meroe Morse, the director of the Black and White Research Lab whom I had met briefly while still in college. She taught me a lot about Polaroid, the power of its developing process, how its film could be 20 times faster (more light sensitive) than other films, and Polaroid's place in the art world, a world she had studied in college. She also introduced me to Polaroid's library collection of great photographs—Adams, Stieglitz, Weston, Strand, Karsh, *et al.* I showed her many of my prints, which she critiqued and encouraged me to display.

Although Dr. Land knew of my engineering creativity, he must have learned of my artistic side from Meroe because soon afterward he began to call on me to undertake various projects with artistic content. One of these was to photograph his younger daughter, Valerie, as a birthday present for his wife. Lucky for me, she was pretty and had an expressive face. The original plan was for a single, artistic photograph, but it ended up with a leather case full of mounted photographs. The collection included original Polaroids, prints made from Polaroid P/N negatives, and prints from 35 mm film: small ones, 8x10s, and 11x14s. All of the printing was done in my kitchen darkroom in long, intensive sessions over three days. A week or so after this project, Meroe asked me if I would like to attend the Ansel Adams Yosemite Workshop. You bet! The workshop covered the first two weeks of June.

The Ansel Adams Workshop was a wonderful experience. Ansel was warm, likable, and very personable in addition to being very intelligent, knowledgeable, and creative. Also, this third visit to Yosemite was made even more special by the personal tours led by a man with many years of first-hand insight and knowledge of the area and history. (See *Yosemite Workshop* below.)

Unfortunately, after the Workshop, I never fully put my heart into photography again. I think the malaise stemmed from two coincident sources occurring that same season. Firstly, the prospect of ever achieving the technical and artistic perfections that I had just witnessed at the Workshop was overwhelming. Secondly, the joy of making photographs was lost after the taste of commercial formality for the intensive portrait project followed by a short-lived contract from the Beacon Press for a book of artistic photographs of plants at the Audubon refuge on Cape Cod.

• • •

After listening to the decade-old radio for about a year, I upgraded my **music system** to include a Fisher tuner-amplifier, a Sony reel-to-reel tape deck, and two AR-4 speakers. My tape collection was largely recorded from the air. I typically listened to classical music every evening at home whether printing pictures, making sculpture, reading, or sometimes recording a tape. Every year while on Hayes Street, I bought a series of tickets to the Boston Symphony Orchestra rehearsals. There were six or eight open rehearsals each year, and I could choose different places to sit for trying different perspectives. I continued to sing with the MIT Choral Society, and those works were always longhair—Verdi's *Requiem*, Beethoven's *Missa Solemnis*, Brahms' *Requiem*, Bach's *Magnificat* and *Mass in B-Minor*, Faure's *Requiem*, etc. In 1964 we had a concert with two shorter pieces: Beethoven's *Choral Fantasy* and Ernst Levy's *Gaudeamus*. The *Fantasy* is the closest that I've ever come to singing the *Ninth Symphony*, and *Gaudeamus* was a premiere that was conducted by Mr. Levy himself. Klaus Liepmann, our director always hired professional orchestras to play for us—sometimes the Boston Symphony Orchestra.

My interest in **classical music** had developed in early high school, and somehow Bill and I encouraged Dad to purchase a hi-fi system and LP records. By the time I left home for good after my sophomore summer, he had an extensive collection of the classics so I was familiar with many pieces.

Music and photography **weren't the only arts** that I enjoyed at that time. I often visited the Museum of Fine Arts and the Gardner Museum.

Not that beekeeping is considered an art, but I approached it as such. Even without much in the way of woodworking tools, I constructed and installed a two-tier observation hive in the bedroom. The upper chamber was separated from the lower by a small queen excluder I made to the required precise dimensions. I had to buy parts and foundations for ten frames just to build two, and the four glass panels were extra thick. There was an entrance ramp that extended under one of the sashes to the outside. I ordered three pounds of bees and picked them up at the Cambridge Post Office—"Oh, yeah, we know where the bees are...come back here yourself to get them." Cardboard covers kept the bees in darkness except when watching, photographing, or showing them to visitors. The hive was active and healthy the first season, and I observed the making of a second queen and the swarming of about half the bees. Wax moths invaded the hive the second season and the hive soon failed.

In 1967-68, I took two successive silversmithing courses at the deCordova Museum in Lincoln. I really enjoyed that and made some nice pieces. Afterward, I bought a collection of unfinished hammers and stakes, but was sidetracked from silver by switching my interest to the apartment house that I bought the following summer.

During those four years on Hayes Street, I drove to New York City two or three times each year for a weekend of art. Destinations included the Museum of Modern Art, the Metropolitan Museum, the Guggenheim Museum, the Huntington Hartford Museum, and others. At the Huntington Hartford on

Columbus Circle, I saw an incredible collection of Salvador Dali paintings and Señor Dali himself. I usually found a cheap hotel for Saturday night, a YMCA, and once in a real flophouse getting the $2.65 "skylight" room, meaning no windows. There is a certain enrichment to slumming occasionally!

• • •

Over the course of my four years on Hayes Street, I built four electronic devices. The first was built from scratch, the others from Heath Kits.

ELECTRONIC PROJECTS – During my first years at Polaroid when still very enthusiastic about photography and beginning to dabble in portraiture, I noted that the portrait studio used lights with both a low-level incandescent lamp for composing and an electronic flash tube for picture-taking. I also discovered a trove of surplus electronic flash tubes and capacitors available for the taking. I combined these two discoveries and undertook making my own **electronic flash** system. A bulletin from GE provided the basic charge and trigger schematics. Except for the flash tubes and capacitors, the rest of the hardware was purchased from Radio Shack and Lafayette Radio including a soldering iron and a collection of color-coded wires. The schematic is marked *AHB 3/66*. The system ended up with four flash channels, 6-pin plugs for each flash, switches for selecting between 100 and 1,000 microfarad capacitors, a perf-board for the charging, triggering, and indicating circuit components, and everything mounted in an aluminum box about 6 by 8 by 10 inches with carry handle and a recessed 120 volt input plug. I had fun making this system, but beyond a few test photographs, I never made a portrait with it.

While making the flash, I occasionally needed a voltmeter, which I didn't have. In those days, voltmeters were galvanometer-based, expensive, and not very good for small-scale electronic circuitry because of the load that the galvanometer itself put on the measurement. So I got a Heath Kit **vacuum tube voltmeter**, known as a VTVM, to begin my collection of serious electronic equipment. Thus, my second electronic project was the VTVM, a relatively simple kit-building project with one vacuum tube, a multiposition switch with lots of precision resistors to be soldered in place, a galvanometer, a hard-to-replace battery, and an aluminum enclosure.

A year or so later, I built a **shortwave radio**. That had two large circuit boards with lots of small components, and a mechanical tuning capacitor with a loop of string and pulleys to move the cursor across the four frequency scales. The radio required a long outside antenna wire stretched horizontally, which I ran from my kitchen window to the top of a fence post at the back corner of the property.

In 1967, it seemed like time for a TV. Again, I built a kit. It was a **black and white TV** set about 11 inches and cost $119 for the unassembled kit. I could have bought the equivalent TV locally for less—but then I wouldn't have had the fun. Building went smoothly, and the picture quality was adequate, but the vertical-hold circuit was terrible. After several minutes of watching, the picture would roll once, then be OK for a minute or so, then roll once or twice, then OK for 30 seconds, then roll a few more times, eventually spinning quite fast. Adjustment of the vertical-hold knob would just start the progressive deterioration over again. Eventually, in disgust, I wrote to Heath Kit and was sent a repair kit for that already-known problem—registering the product didn't help that time. It was much better after the repair, but never perfect.

But did I have a **social life**? Oh yeah! This fenced-in compound, sometimes locally called *Pilgrim Village*, was largely occupied by young, single, working people, some of whom I got to know, and one of whom I dated for a while. But Village parties were sometimes a resource for meeting people from outside the compound. Polaroid was a mother lode of fascinating people, both male and female, and of the latter there were many young and attractive women that I got to know, some of whom I dated and one that I married. One engineering colleague, Bob and his family became life-long friends. Another, Len and his wife Ruth became occasional friends although they were a generation older. Doug, a work colleague and his wife Kay introduced me to a large circle of mostly Lexington friends. Sue, a former girlfriend that I had met junior year was then living in Cambridge and we renewed our friendship on an occasional basis. I got to know several families and single women through the MIT Choral Society. For two of the BSO rehearsal seasons I bought two sets of tickets for taking dates.

Parties, parties, I threw parties. Although that apartment was pretty small, I could cram in a real passel of people. Having a bent for statistics, I often kept notes of food and liquor served and left behind to improve my stocking efforts for the next time. For a few examples from the notes still in my archives: in November 1966 I invited all 18 work colleagues, all of whom showed up with their wives for a total of 37 crammed into that two-room space. In November 1967, there were 21 friends over, and my notes estimated its cost at $16.65 for liquor, soda, shrimp, cheese, and crackers. In May of 1968, I had 16 people for cocktails ($11.78) and dinner (about $25) including 11 pounds of roast beef.

In September 1968, the day before moving out and not long before the building was to be demolished; I threw a painting party where people could paint anything on the walls, windows, ceilings, anywhere. I had water-soluble paint solely for the welfare of people's clothing.

The next day, I left the building with the last of my clothes and the kitchen sink.

THE POLAROID YEARS
December 14, 1964 to March 1971 – Ages 24 to 31

I first saw Dr. Land[6] at an end-of-year rally in the cafeteria downstairs from my office at 28 Osborn Street, Cambridge. I was 24 and had been working at Polaroid for only about two weeks at the time. His visage was familiar to me from years of interest in photography and reading about Polaroid and its founder in *Popular Photography* and other places. I was thrilled.

I had no further contact with Dr. Land for nearly a year. I worked on a variety of miscellaneous projects that first year including a stereo camera and viewer, factors affecting redeye, automatic focusing, a shutter for the ColorPack camera, a self-erecting bellows (US Pat. No. 3,418,907 and back cover upper left), and a pocket-sized camera incorporating mirrors. Once, when discussing the sketches of the mirror-camera with my manager, Dick Wareham, he commented, "You seem to understand mirrors pretty well." I replied something like, "I've liked optics since my youth, and both my high school Michelson interferometer and my bachelor's thesis included mirrors."

Shortly thereafter in late October 1965, Wareham came to my office with a big grin on his face and said, "I've got something for you that's right up your alley." We went to his office where he showed me a Cross pen and pencil case and said, "Land wants a camera that is this size and doesn't need to be opened up to take pictures." He went on to describe the concept of a **scanning camera**. The pen case measured 3.50 by 6.75 by 0.87 inches with well-rounded edges and corners, and the top surface slightly domed and thinner towards the ends and corners—thus it slid in and out of a jacket pocket or woman's purse quite easily.

Overnight, Wareham drew one of his inimitable perspective sketches of his vision of the basic working parts within the camera (similar to Fig. 1 in US Pat. No. 3,468,229). The plan was to hold the camera vertically and flat against one's face—like peeping over the top edge of a book. The sketch showed a viewfinder and rangefinder in the upper left corner, and a centered window near the top outfitted with a 45-degree mirror mounted on a transverse axis. The mirror would scan over the intended scene from top to bottom and project the light downward toward the lens, through the lens, and downward to a very elongated mirror mounted at 45 degrees just in front of a narrow slit where the film would slide past during synchronized scanning. A vague film box hovered in front of the light-cone, and there were spread rollers and drive wheels near

---

6 Edwin H. Land (1909-1991) was founder and president of Polaroid Corporation. After one year of chemistry at Harvard, he dropped out to form a business around his early invention to create sheet polarizers quickly and inexpensively. The name Polaroid was coined from *polarized* light. All their cameras were branded Polaroid Land Camera. Land was brilliant, and could concentrate his mind to focus on a needed invention, for which he ultimately had 535 US Patents awarded covering many areas of technology.

the bottom of the camera. There were also a lot of omitted details, so very many of which we would learn about over the next 20 months.

It was my assignment to explore and demonstrate these possibilities. Wareham had no experience with a scanning camera nor its potential complexity, but I was intrigued by the concept and confident that I could work through the optics. As an experimentalist, my first thought was how could we prove-out the basics of such an approach without also designing and constructing the entire miniature camera?

Within a day or two, I developed a plan for **building a prototype**. The optics and mechanism for articulating the optical components was built on the front of a Model 100-style Polaroid Land Camera back. Although such an approach would never meet the size requirements, it should give us much functional experience. This proved to be either triumphal or disastrous, depending on one's level of cynicism. It was triumphal from the perspective that as we learned more detail about scanning optics, the underlying structure permitted us to easily modify the mechanism to refine its optical functionality. It was disastrous from the perspective that our success with this prototype deluded us, or rather, it deluded Land, into thinking that a scanning camera had a real chance to become a commercial product. (This scanning camera built upon a Model 100 chassis was soon named LEN'S & BELLOWS' CAMERA and will be referred to as such to distinguish it from the several other meanings of "SX-70." This camera was intended solely as an experimental test-bed—never as a product itself.)

I quickly set about designing this revolutionary new camera. As the design took form, Wareham arranged for one of the best model makers in our shop to build the hardware—the Len of "Len's & Bellows'." The project moved so efficiently and rapidly that, miraculously, we were taking pictures with a scanning camera in less than six weeks. Wareham was so pleased with the results that when Land asked him how the camera was progressing, he invited me to go with him to present a functional demonstration. What excitement. Land was excited, too, and over the course of the meeting managed to figure out which way he and Bill McCune should respectively walk to make himself look thinner and Bill, the health and exercise zealot, look portly. (Scanning cameras distort any objects that move during the scan.)

That first meeting with Land was very enlightening for me. Although I had known of him for over a decade and had heard many people speak of him in awe, I quickly discovered that he put me at ease and made me feel very self-confident in his presence. He was very perspicacious and articulate, thus fascinating to be with and listen to. On many occasions we built quickly on each other's ideas and muttered musings.

There were several more such technical meetings with Land over the following months, some with me alone. He dubbed this camera project SX-70 and explained that during WWII when Polaroid had a number of military contracts, the project numbers comprised the two letters "SX" followed by two digits. The picture in a minute project fell under "SX-70, Miscellaneous" He was using the same project number for sentimental reasons. No one ever expected the name to survive into the marketplace.

• • •

Before delving into Polaroid cameras, especially in this now-digital age of photography, a brief history of camera and film functionality might be useful.

PRE-DIGITAL PHOTOGRAPHY – For a century and a half, photographic images were captured by having a lens focus an image of the world onto a transparent sheet of glass or plastic coated with an emulsion of silver halide. A brief exposure of light triggered an invisible change in the silver halide molecule. When that exposed emulsion was put into a solution of "developer," the light-struck regions of the emulsion precipitated onto the glass or plastic sheet as metallic silver in a microscopic sponge-like form that appeared black. The adjacent regions that were not struck by light did not precipitate, and the remaining silver halide was subsequently dissolved away when the sheet was moved from the developer to the "fixer." (Daguerreotypes were a variation on this theme.)

After washing and drying, the sheet's image could be seen with back-lighting to be black where the world was white, clear where the world was black, and a host of gray shades in between. Thus it was a negative image, and this sheet was called a *negative*.

To make a viewable print, the process was repeated in which the negative was back-illuminated with a light bulb, the image projected through the lens of an enlarger, and projected onto a piece of white paper coated with a similar type of emulsion used for the negative. The piece of paper was then developed, fixed, and dried. Thus another negative was made, but two negatives make a positive, and the result was a black and white photograph.

(Color film was complicated by having multiple layers of silver halide sensitized to the three primary colors and later linked to dyes to form a color image. Otherwise, the process was similar.)

The magic of the Polaroid instant picture process was invented by Land in the early 1940s and first marketed in 1948. Rather than make prints in two steps, the Polaroid process placed the white print paper (positive sheet) against the undeveloped film and squeezed a thin layer of developer between them. This very specialized developer precipitated silver onto the negative as described above, but the unexposed silver halide diffused through the thin layer of developer where it was triggered by a chemical coating on the positive sheet to precipitate as black on that positive sheet. After a short period of time (one minute or less), the two sheets were peeled apart, the negative discarded, and the white sheet viewed as a positive black and white photo. This is known as the diffusion transfer process. (Color pictures were similarly formed, but the chemistry of the emulsion, developer, and positive coating were far more complex.) This system resulted in a print the same size as the negative, with the requirement that Polaroid Land cameras needed to be large.

• • •

To roll back this early scanning camera narrative and explore the project in more detail, I'll return to the point of embarking on a test-camera to be built on the 100-style back.

My major concern regarded the potential **complexity of a scanning optical system**. I knew enough about optics to recognize that a simple scanning system would produce considerable barrel-like distortion, that its correction would be difficult at best, that holding a camera steady for a third of a second was expecting a lot, and what about flash? In fact, I was so skeptical of this project that I proposed an optical test-bed in which not only could the optics be changed-out with relative ease, but that we need not waste time designing a complete camera with film-scanning and developing features just to dispense with this impractical idea—hence my reason for utilizing the 100-style back for the film system. I did not share my skepticism widely. Despite my doubts, I was intent on doing the best job possible, which explains my choice for the first, distortion-free optical system as shown in the illustration below rather than the simpler, barrel-distortion system expected of Wareham's sketch.

TEST-BED PLAN – The Model 100 Polaroid Land Camera back appeared to be the ideal choice. There is one step in the use of that type camera in which the negative translates across the back, such as would be needed for a scanning optical system. Looking at the details of the Model 100-style film pack, the negative faces forward toward the rear of the camera's lens for taking a picture. After the picture is snapped, the white-tab pulls the negative across the film plane, around the end of the film pack, and into a position where it faces rearward directly toward, but in loose contact with the mating positive sheet. Once the white-tab has been fully extracted and pulled free, a yellow tab appears. Pulling the yellow tab results in pulling the negative, the developer pod, and the positive sheet through the spread rolls and out of the camera where it develops. After 60 seconds, the user pulls the photo apart from the negative. The negative, excess developer and two tabs are discarded. (Black and White film required only about 15 seconds to develop.)

Therefore, if the scanning optics could be synchronized with the process of pulling the white tab, a scanning camera could be built. Synchronization between the film and a moving or scanned image had been demonstrated in panoramic cameras invented in the late 1800s, in horse racing's photo-finish cameras, and in the Xerox copier introduced in 1959.

The basic design approach for that scanning camera was mine—use the type 100 camera back for the film-handling, outfit it with a light-tight barrier over the film plane except for a narrow scanning slit at the left end of the film gate, install an interchangeable optical system in front of the slit, and translate the film past the slit by pulling the white-tab, all while driving the optical system in coordination with the pulling of the white-tab. The coordination was to be accomplished with a frame-like structure surrounding the camera back in which the top and bottom members of the frame could slide right and left in tracks added to the camera back, and in which the frame was long enough that it would slide about 4 inches, a little more than the length of the film. In use, the frame would start in its left-most position, the white-tab would be clamped to the right end of the frame, the trigger would be released, and the frame would move left to right. Linear cams screwed to the top and bottom members of the frame would drive the optical system at the same time and rate that the film was being pulled across the exposure slit by the white-tab. The slit was about $1/4$ inch wide and adjustable. Since there was no shutter at the lens, light from the world was always falling upon the slit. As a result, the developed print had a $1/4$ inch wide white band at its right edge. Details of each cam were determined by the linkage from the cam follower to the respective scanning mirror or lens. (See illustration on a later page.)

As the design began, Len, the previously mentioned master model maker was assigned to this project. He was ideal—he could make anything. We got along well, and I quickly learned that he could be depended upon to "design in the round." Thus, I could attend to the critical, innovative optical challenges while leaving many of the functional challenges to him. In fact, Len and I made a pact that we would share in the design of sticky details. Whenever we were stumped, we would each think about a possible solution and would use the design of whoever had a good idea first. Also, Len combined our two names to facetiously call it the *Len's & Bellows' Camera*. By avocation, He was a watchmaker, and his workmanship was of meticulous quality.

LEN'S & BELLOWS' CAMERA DETAILS – The linear cams attached to that translating frame were aluminum about 5 inches long, $1/8$ inch thick, and about one inch wide before cutting the curved functional surface of the cam. In the first optical embodiment, there were two cams: one rotated the scanning mirror to sweep the image of the outside world onto the film as it was pulled across behind the film slit; the second translated the lens carriage to maintain the proper lens-to-film distance at all times. This was 1965. Computers were rarely available to engineering projects, and there certainly was no CAD software for drafting on a computer. The cam design and construction was done by hand using a drafting pantograph. The drafting paper I was using was prone to changing size overnight with humidity changes, so it was my practice to schedule a design to be completed as quickly as possible, and certainly within a single day. I drew the critical optical geometry and components four times actual size and traced the rays as meticulously as possible, constructed the corresponding mirror and cam follower angles as accurately as possible, and measured the follower offsets, added the frame translation to the X-value and entered the result into a table of X, Y coordinates for Len to use on a milling machine to cut the cams. These measurements were repeated about 40 times to step along the 3.75 inch length of the types 107 and 108 films.

At the time, there were no numerical-controlled milling machines at Polaroid so Len used the best tool and die-making techniques. After making the cam blank, he milled each of the tabulated cam positions I had supplied, then carefully filed the cam profile tangent to the milled scallops by eye.

I was concerned that all this hand work on both our parts might introduce so many small, cumulative errors that the resulting photographs might have noticeably smeared images or distortion.

We proceeded to assemble and test the camera despite my worries. To my relief, I could not detect any image error. In fact the photographs of the resolution test panel revealed that the resolution in lines per mm was as high as this film and lens were capable of. Both the draftsmanship and the cam construction had been executed with precision.

The **first optical system** used a lens from the Polaroid Swinger Land Camera. This single element "reverse landscape" lens had its small f/17 aperture behind the lens, which simplified the design of the scanning geometry. The lens was oriented to face the subject and mounted on a translatable carriage. A rotating mirror was placed behind the lens and coincident with the aperture. An elongated film-mirror was fixed at 45° directly in front of the film-scanning slit. Thus the light path started with the subject being photographed, progressed toward the camera, through the lens, reflected by the scanning mirror in a direction parallel to the film, traveled about 3.5 inches to the film-mirror where it was reflected perpendicular to the film and through the film slit. The lens-carriage moved back and forth laterally about half an inch to account for the fact that lenses are farther from the edges of the film plane than from the center. This optical system was quite simple: the scanning mirror pivoted around its own axis, the film-mirror was fixed in place, and the lens translated right-to-left-to-right. Two cams were needed: one for the mirror, one for the lens translation. The entire optical path needed to be substantially light-tight, requiring that the pivot shafts and translating components had to be mounted within and accessed through light barriers.

This was the camera and optical system that was working adequately within six weeks to demonstrate in Land's office.[7] Without the handles and drive motor, it took two of us to take pictures—one to hold the camera steady while perched on the edge of the table, the other to pull the frame at a near-uniform rate. Land was delighted. (That day was my first anniversary at Polaroid—the meeting was a monumental present!)

Our next step was to **add a motor** so that it could be used for taking real, hand-held pictures, and then add the peripherals such as handles, exposure release, and viewfinder. The motor was an unusual experience in itself. Sometime in previous years, I had learned about "Neg'ator" springs. These were curled ribbons of steel that produced a nearly constant torque when rolled from a reverse roll to their natural roll. The sample purchased was small enough to be installed within the battery chamber at the left end of the 100-style back. Len used a tiny cable wrapped around the Neg'ator capstan to pull the frame across the camera. But this system needed speed control to make the exposure occur over a third of a second. I called several potential sources for speed governors, none of which were very encouraging: "smallest is six inches in diameter," "will design a custom unit for you but we need a detailed specification and six months," *etc.* While dialing one of the calls, it occurred to me that the rotary telephone dial must have a governor. I carefully opened my phone and looked: sure enough, it was about the right size, too. When no one was looking, I went into an unoccupied office, and removed the dial from the phone and delivered it to Len. Within a few days, he had the whole governed spring-powered motor installed in the camera.

Len added two handles from a Polaroid close-up camera kit and an Albada projected-frame viewfinder Leica accessory. He made a thumb trigger to release the cable-tensioned frame to begin the $1/3$ second exposure. Cocking was done by pushing the frame the other way by hand.

Ultimately, we built about six different optical systems for the Len's & Bellows' camera. The first, described above, resulted in smearing of the image near the corners as it passed the scanning slit. Although subtle, it was noticeable.

> **LEN'S & BELLOWS' VARIANTS** – The **second optical system** corrected this smearing problem by installing a third cam to pivot and translate the film-mirror, but this required an additional, superimposed translation of the lens to maintain the proper lens-film spacing at all times—hence it required a redesigned cam for the lens.

---

[7] Dr. Land's office was unexpectedly simple. It was located at or within a few feet of where Alexander Graham Bell made his first telephone call to Watson. I never heard Land mention that fact, but it must have given him some sense of connection with that other inventor. It was a modest office, especially considering that he was the founder and president of a very successful Fortune 500 company. It was about 14 by 22 feet with three ancient sash windows finished in frosted glass and hardware cloth on the outside for security. Bookshelves lined the other three walls, and three large photographs by Ansel Adams hung on the tips of the shelf brackets such that they could be moved elsewhere to access any temporarily hidden books. Beneath the windows was a desk-height counter for five or six telephones, one of which was red, said to be for calls from the White House. Other furniture included his oversize but ordinary desk, a recliner, and a conference table with about six chairs. One doorway led into his laboratory. Even after the so-called *block house* at 549 Tech Square was occupied by Polaroid executives, Land much preferred the old office and spent little time in his finer office.

This arrangement simulated all of the optical relationships typical of a standard camera, and therefore had no image distortion. However, the f/17 Swinger lens was not "fast" enough for color pictures. (This configuration is illustrated in the figure below.)

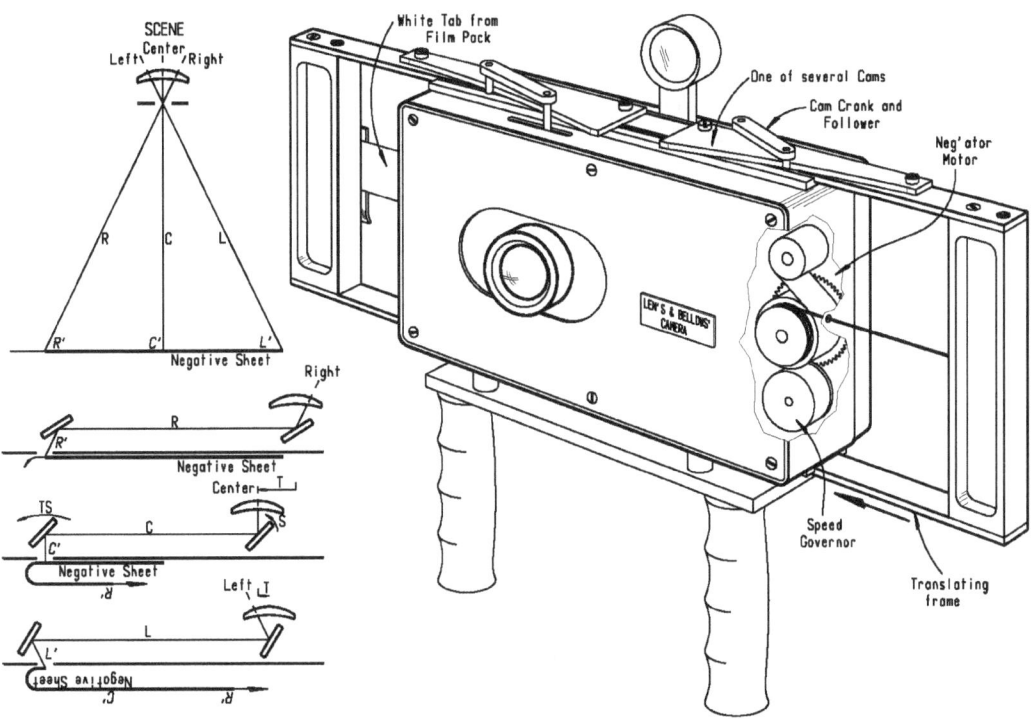

The LEN'S & BELLOWS' CAMERA, a scanning camera constructed in 1964 and '65 and drawn in perspective from memory in 2012. Inset at left shows a conventional camera's optical system followed by three progressions below of the scanning system as it successively simulates the conventional optics. For scale, the "box" portion of the camera measures 7.5 by 4.4 by 1.8 inches.

Therefore, the **third optical system** was designed and installed. It was this system that was used for much of the demonstration and testing over the next many months. This optical system used a much faster f/4.5 three-element achromat lens that allowed us to take color pictures. Again, the standard photographic geometric relationship between the lens and a flat film plane was preserved at all times. The film-mirror had to pivot and translate to reproduce the angle of incidence as the light struck the film, the lens-mirror had to pivot *and* translate relative to the lens to collect the light exiting from the lens at different angles (unlike the Swinger lens), and the lens and lens-mirror subassembly had to translate to reproduce the path length from lens to film for each position along the scan. These overlapping functions needed to be synchronized and isolated from one another, resulting in a complexity of cams and cam followers. We readily recognized that such a complex optical system could never be practically employed for a compact camera.

Therefore, we felt it necessary to determine if a much simpler scanning optical system might be acceptable. This configuration would be virtually identical to the scanning camera plan originally described by Land and sketched by Wareham in October 1965. For this **fourth system**, the lens was placed close to the masked film plane and pointed parallel to the film plane and toward the right of the camera. The scanning mirror was to the right of the lens (in front, optically). This optical system was quite simple: the mirror pivoted around its own axis to scan the scene right to left, and the lens and film-mirror were fixed in place. Only one cam was needed to rotate the scanning mirror.

This system was already anticipated to cause a significant amount of one-dimensional barrel distortion—our objective was to determine if such distortion might be acceptable.

Used in its most conveniently-held position, the Len's & Bellows' camera scanned right to left. This fourth optical arrangement caused straight horizontal lines near the top and bottom of the picture to bow away from the center of the picture. Usually, this effect was not too distracting, but Land's idea for a product would require the scanning direction to progress from top to bottom. It was awkward to hold the Len's & Bellows' camera in this position, but we did it. The resulting pictures distorted straight vertical lines near the edges of the picture to bow away from the center. It was surprising how many pictures had such lines—doorways, buildings, tree trunks, even people—and how distracting the results were. Group pictures of people showed those standing at the right or left to have very bad posture. (Two of my patents cover inventions to eliminate this "cylindrical perspective distortion." See US Pat. Nos. 3,468,229 and 3,468,230)

Edwin H. Land at the time that I knew and worked with him. Land was the founder, president, and technical leader of Polaroid. I was blessed to have worked with this extraordinary person. (Photo by Jay Scarpetti.)

I never called him by any name other than *Doctor Land* although around the office, we usually referred to him as *Land*. He was never called *Ed*, but his friends and close associates called him *Din*. The *Doctor* title was from an honorary degree.

One Saturday morning in March 1966, Land called me and invited me to breakfast at the Parker House, a landmark old-world hotel in Boston. He wanted to talk about the ideal lens for the scanning camera. This was my first experience observing him invent something. He was truly able to concentrate on the job of stimulating creativity both of his brain and mine. In retrospect, I was amazed at my own level of contribution during that hour or two that we talked and drew on scraps of paper. When we finished, we had the outline of a lens specification that might serve for a simple scanning system, yet eliminate or reduce the distortions inherent to the simplest optical system. He called it the "Parker House lens."

Incidentally, this was the first of many times that Land called me at odd hours to talk at length or invite me to a meeting, a meal, his office, someone's lab, or his home to discuss camera issues or, occasionally to be sociable. The latter included a small dinner party with his family and Ansel Adams[8] and a larger cocktail party where I met Harvard's Prof. Edward Purcell a Nobel Prizewinner among other impressive intellectuals. He also encouraged and paid for me to go to the Ansel Adams Yosemite Photography Workshop. This habit of his for bypassing the usual management structure was not uncommon at Polaroid, and my boss, his boss, and the vice presidents above them were used to this breaking of hierarchy and not offended by it.

• • •

In August 1966, our small design team of one engineer (me), three designer/drafters, and a model maker was suddenly enlarged to a project total of 19 and moved across Main Street to the ninth floor at 565 Tech Square. Dick Wareham was the manager, I was the senior engineer, and there were 2 junior engineers, 4 designers, 6 drafters, and 5 model makers including Len with a well-equipped model shop. We occupied half of the ninth floor very sparsely while the other half was kept vacant. Land wanted the project to be secret and no one beyond him and the 19 of us were allowed in—not even our secretary or any of the vice presidents.

About that same time, Land hired James G. Baker as a consultant to design the Parker House lens for us. Dr. Baker had designed reconnaissance lenses and cameras during WWII and for the U-2 spy plane project on which he and Land had worked together. Land described Baker as "vastly imaginative." Baker, Wareham, and I met at least twice a month for several years, and Land sometimes joined us. During the first year, all the effort was on the scanning optics. The Parker House lens, *per se*, was never designed as its conceptual outline did not lend itself to practice. But Baker worked out numerous designs to compensate for the many distortions, all of which were complex and most of which included some combination of aspheric surfaces, counter-rotating prisms, waggle (or Waughghoul) prisms, zoom lenses, tilting lens elements, pentaprisms, and even a twisting mirror. He would show up with a 2-inch stack of IBM tractor-feed paper that included the calculations of his most recent design. None of these designs satisfied Baker's criterion of performance nor our criteria for reasonably fitting into a compact camera. In time, several dozen optical ideas were proposed by Land, Baker, Wareham, me, or others including two of the patent attorneys. These ideas included miniversions of the Len's & Bellows' optics, folded duplets, collapsing triplets, zoom lenses, concave toroidal mirrors,

---

8 An example of Ansel's humor: At that dinner party, Mrs. Land happened to drop a gooey hors d'oeuvre on the part of her dress that covered the upper slope of her breast. Ansel quickly jumped up with a big grin, grabbed a napkin and said "Let me clean that." Mrs. Land responded with a laugh "Get away from me you naughty man!"

etc. in combination with two, three, or five mirrors in various configurations of orientation, compactness, or translation during exposure.

Hundreds of pictures were made with that clunky Len's & Bellows' camera. But the **scanning camera had serious flaws** that would ultimately kill it. The full exposure took about a third of a second. It was neither easy nor intuitive to hold a camera absolutely steady for that length of time. The instantaneous shutter speed (exposure through the slit) was fast enough to eliminate most blur, but the shape of the image became seriously flawed if the camera had any gross motion during the $1/3$-second picture taking. That long exposure period also created a serious challenge for a flash bulb. There was no possibility of making time exposures or even using somewhat slow shutter speeds. Another problem was that pictures taken under fluorescent lamps resolved the 120 Hertz flicker of the light-producing arc and caused closely-spaced light and dark bands across the picture. With meticulous adjustment of the slit width, it could be minimized but never completely eliminated. (Note: Ten samples of a slow-burning flash bulb were custom made for us to experiment with, but they tended to flicker as the flash progressed and were so dim that we had to mount them in a scanning reflector that followed the portion of the scene being scanned at every instant.)

All told, we worked on scanning cameras from November 1965 until June 1967. Although Dr. Baker was a very imaginative lens designer, he never solved all the optical challenges in a practical design.

CHALLENGES – None of our efforts at designing working mechanisms, even while ignoring the lens complexity, ever approached the size goal that Land was trying to achieve by using a scanning approach. In that entire period, we never built a single "camera" because we never had enough of the design details to even envision an assembly. We did, however, make numerous fixtures to test individual systems and assemblies. Early on it was decided to have a square picture so that the user would not need to reorient the camera to an awkward position. We didn't know about the integral film of the future so all our early work assumed a variant of the peel-apart pack film. A number of fixtures tested various methods of scanning the negative, then combining it with the positive, then effecting a spread. To highlight this one challenge, a 3-inch picture with borders and a scanning leader would be about 4 inches in total length, would need to travel about 4 inches just for the scan, then need to travel another 4 inches for the developer spread and capture into a dark chamber. Fitting such a 12-inch film-transit path into a 5 or 6 inch long camera was very cumbersome, and none of our efforts worked reliably. Even the ones with fold-out or trombone extensions worked poorly.

Other experiments included wind-up motors with speed governors, miniature motors and gear trains, scanning drive systems with tabs or belts, small diameter spread systems, spread rollers with internal spring- and electric-motors, assembly structures for the negative and positive, and others that I have long since forgotten. (Years later when I was deposed in the patent infringement case against Kodak, I could not believe the numerous cameras and experimental fixtures arrayed as "exhibits" over five or six table tops that were attributable to my design work or overview.) When the integral film with its reversed image came to our attention in the fall of 1966, Baker's optical system needed a third or fifth mirror, which forced the optical system alone to nearly fill the camera's thickness.

It was a wonder that this team of 19 people could maintain even a modicum of enthusiasm, but it did. For me, at least some challenge was provided by designing and building several alternate configurations of super-small, but folding cameras with and without mirrors. I called these SX-72 cameras and had Land's approval to explore them but not his interest (see one example in Fig. 8 of US Pat. No. 3,683,770 and back cover lower right). Also, at various times, I was engaged to design Dr. Land's table for the new Board of Directors meeting room—elliptical; serve as liaison between Polaroid and its consultant Ansel Adams; design a shutter and electronic circuit for what Land called the "Homebody" camera (see US Pat. No. 3,498,194 and front cover upper right); serve as president of the Photography Club; attend the Ansel Adams Yosemite Photography Workshop for two weeks; take a lens design course at UCLA; and take several days out to make dozens of photographs of Dr. Land's daughter.

The **issue of mirrors** brings up an interesting point. During one period, Land sometimes included mirrors in his discussion, ostensibly for the purpose of compactness. Whenever I pointed out that a single mirror or three mirrors would result in a reversed picture, he would always acknowledge the problem, spend a few moments fretting over it, then suggest that we should "not try to resolve all details today," and then go back to a mirror discussion that included an odd-count of mirrors.

At the time, we had no idea that his film chemists were designing a front-exposure film that would result in a mirror image. Presumably, that group was also warning Land that such a film could not be sold because of the mirror image. He kept the two groups apart for many months before we were introduced to one another with the dawning realization that the two useless products, when used together, produced a useful product. In retrospect, how was he able to inspire us and keep our two groups working diligently on products that we each thought could not be sold? He was an amazing man.

• • •

In June 1967, after everybody, apparently including Land, had become discouraged with lack of progress and mounting awareness of major obstacles, Land went on vacation to California with his wife. Even on vacation, his real interest was Polaroid, and I got many calls from him to discuss alternate approaches. During one of those calls he proposed a design in which the optical path was folded at its midpoint by a relatively large mirror and the lens tucked down close to the spread rolls. The mirror-holding panel and shutter panel were part of a four-bar linkage that would collapse flat to resemble his original objective of a flat camera that would fit in a pocket or purse. Wareham had been following our discussion on the extension phone, and when Land and I finished the call, he came out of his office, gave me a high sign, then came over to my desk and said, "Finally, he heard you!"

After these calls, it was my follow-up job to make sketches and to critique the idea. This particular idea was no exception, so I went to work. And the full-scale layout worked. **Thus was born the SX-70** as it eventually came to market. (US Pat. No. 3,683,770 to Land and Bellows was filed a few weeks later, though it took years to issue.)

Our shop made a wood and sheet aluminum model just to see the shape and folding action. It was airmailed to Land in California.

We were able to build a working camera fairly soon, but it was a solid nonfolding camera, and we used an opaque cardboard box to "muzzle" over the front of the camera before ejecting the film. The box would serve as a darkroom in which the film could develop for a couple of minutes before exposing it to the light and taking a peek. The color lab had not yet figured out how to make the developer self-opaquing, thus the need for darkness.

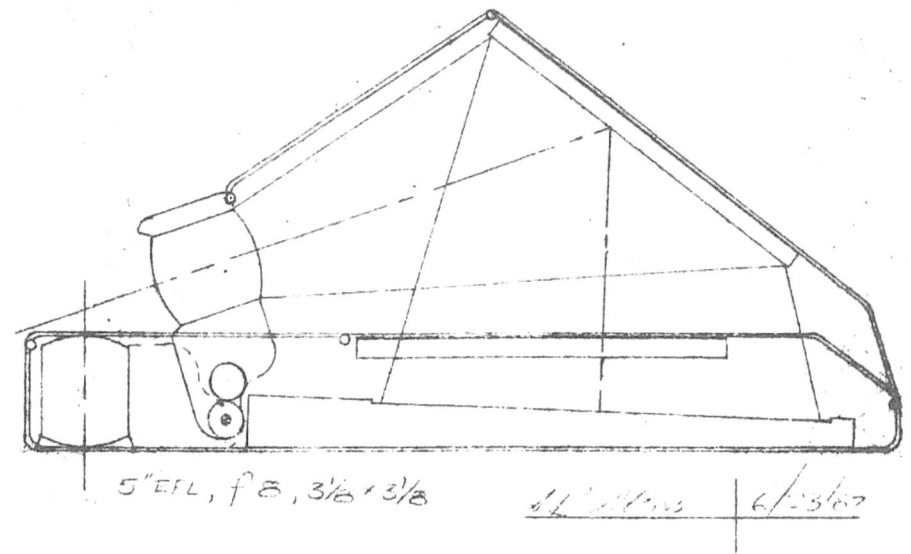

My original drawing of the SX-70, June 23, 1967. Reproduced here at 75%.

(The story told to me by Wareham about the evolution of opaque developer is another window on the intense brilliance of Dr. Land. After many months without progress by the color lab, Land spent a concentrated weekend alone inventing potential chemical mechanisms to make the developer opaque immediately after spreading but still allow the image to emerge a few minutes later. On Monday morning he presented and discussed his list of 24 different possible mechanisms with the color chemists, and together they selected eight of the most promising. The lab experimented with those eight for about a week, and then selected three as highly promising. Land called in the carpenter shop to partition the lab into three sections and formed three teams to engage in isolated competition. In about six weeks, a clear winner had emerged and the walls were torn down. I never heard any confirmation of this story.)

Even the solid version of this collapsing camera left many issues unanswered. The SX-70's awkward shape and folding linkage made viewfinders and rangefinders difficult. I, and later Phil, a young engineer who was working for me, built or designed over a dozen different versions, and there were many more concepts sketched on paper (for patented examples of mine see US Pat.

Nos. 3,554,076 and back cover center left; 3,622,242; 3,680,946; 3,610,123 and back cover upper right; 3,610,128; and 3,619,202). The folded light path that then collapsed flat presented a challenge for a fold-up light-tight bellows. One solution used overlapping and interlocking leaves on each side of the camera (illustrated in Figs. 1 and 3 of the basic patent No. 3,683,770). Ultimately, the design settled on a rubber boot that could stretch and accept any contortion needed as it folded open or closed.

Early attempts at incorporating a motor of standard or custom design were fruitless because of their relatively large size. One of our engineers with young children came into work one day with a slot-car and the observation that its motor was both strong and small. He sought out the manufacturer and contracted him to custom-make a similarly effective motor for us. An early idea of mine was to step down its very high speed with a high-reduction Harmonic Drive™, and one of my patents incorporates such an arrangement partially within a spread roll (see US Pat. No. 3,505,943 label number 58 illustrated as a Harmonic Drive). Ultimately, it became more efficacious to put the motor at the back of the camera and transmit the rotation to the front by means of many gears, each of which reduced the speed in small steps.

Several shutter designs were developed; some with the expectation that they would be controlled electronically and cocked by the motor (see US Pat. Nos. 3,618,501; 3,545,352; and 3,557,678). One of my designs was a flash array with ten nesting bulbs and reflectors to equal the number of pictures in a pack (see US Pat. No. 3,614,412). The original squarish array was later rearranged to match the width of the bar-style viewfinder and, even later, the width of the shutter housing. That style was marketed as the "flash bar."

One of these fixtures was the subject of what we sometimes called **Landisms**: brief, succinct, pithy quips that Land offered, frequently with humor. I was making a spring-driven spread system similar to a spring-driven window shade. The fixture was unfinished, but the oversized hollow spread roll, its long floppy spring inside, and a windup crank at one end was lying on the bench when Land showed up one day asking, "What's new?" I showed him this. The thin-walled satin-finished roll was about $3/4$ inch in diameter. Land cradled it in his hand, wound up the crank, and slowly let it unwind within his grasp. At one point he suddenly looked at me in surprise and exclaimed, "I can feel the spring squirming inside the roll." I frowned and said, "No, that's just your imagination." He then handed it to me to try, which I did as he said with a twinkle in his eye, "Over the years I have learned that my imagination is a true statement of an unobserved fact." Just as he reached the word *fact*, I began to feel the spring squirming for myself. That was a distinctly well-timed Landism.

I always wished I had recorded those many Landisms I heard or heard about during my tenure there. Two of my favorites are: "Nature is on your side," and "An unsuccessful experiment is not a *failure*, it is a confirmation of something you cannot do."

• • •

Although unrelated to the design of cameras, in the fall of 1967 I was summoned to meet with Polaroid's CFO. He had the surprising news that I was being awarded a **stock option** for 400 shares of Polaroid. Not only did he explain the nature of an option, but he told me that at Polaroid, a stock option was one of Land's methods of personally rewarding exceptional employees. I had stars in my eyes, was overwhelmed, and graciously thanked Dr. Land next time we met. I undertook to set aside thrice my salary in order to exercise those options five years hence. At the time, the stock had been doubling every two years, so I was already feeling rich beyond belief! But, alas, the growth in PRD stock value soon flattened and fluctuated without much spunk.

The author in the Tech Square offices of Polaroid in about 1969. (Taken with a Model 100 Polaroid Land Camera, not an SX-70.)

Not only is engineering a good profession, but it can provide a good lifestyle. With discipline, an engineer can meet the family's needs and still set aside a portion of the paycheck for investment. Paying for the stock option was well beyond my current savings, but in one form or another I expected to be able to exercise most or all of it. However, a year later when I chose to buy an apartment house, I used much of the savings thus far to buy it with the reasoning that owning and improving real estate was also a good investment. Ultimately, the apartment house earned a huge return while the leveraged stock tumbled before I finally sold it years later at a small loss. Not all opportunities are winners, but if the opportunities can't be taken, there is no winning.

• • •

The integral film was revealed to us during development of the scanning version of SX-70. In its earliest form, the positive and negative sheets were assembled as two squares with the developer pod taped to one end, leaving the pod comparatively floppy and disfigured with the typical pattern of lofting and sealing marks. By installing the film in its most natural position in the not-yet designed scanning camera, the pod ended up at the top of the picture. We never had a good solution for this unfortunate arrangement, but as I was designing the folding version of the SX-70 that day over the phone with Land, I noted that the pod would be at the bottom of the picture where it would be more aesthetic and could serve as a space for a caption. Soon afterward, I suggested to the film lab that the positive sheet should be rectangular and extend fully over the pod, thereby making that area stiff and uniform in appearance. Thus was born the format, still symbolically seen today in movies, advertisements, cell phones, *etc.*, of printed **photos with a wide border at the bottom**—just as deckle edges were once symbolic of photos up through the 1950s.

     Dr. Baker shifted to designing a lens for this camera. This was no easy task either.

> **CLOSE FOCUS** – In the early months of tuning up the folding structure, the focal length of the lens changed several times; the relatively short focal length settled upon required a challengingly wide field; the lens had to be quite thin; and the unique demands for the front-element focus were challenging. By this time, Louis Alvarez, a Nobel prizewinner friend, had given Land a pair of clear plastic plates with one aspheric optical surface that when turned backwards and translated against one another formed a continuously variable lens ranging from –2 to +2 diopters. Such an adapter in front of a "nearsighted" camera lens would provide focus to within 10 inches of the camera—a nearly half scale image in our case. With this carrot in mind, Land asked Baker to design a front-element focus lens for the SX-70 that could focus from infinity to the distance needed for half scale: about 9 inches. Such focal range was unprecedented in front-element focus lenses, but Baker achieved it. (A variant of the Alvarez optic was incorporated as the "quintic" focusing system in the much-later Polaroid Spectra Land Camera, a variant of the SX-70.)

     During this period, I continued to be the guiding engineer on the project overseeing all aspects of its design. This was an informal position—I was never appointed as chief engineer or project manager. In fact, Wareham continued to take a day-to-day interest in the entire project, but he had so many other balls in the air that the major load of overseeing the other engineers, designers, drafters, and model makers fell to me. Wareham once gave me a high compliment: "When you work with others, including draftsmen and technicians, you don't *tell* them what to do, you *teach* them what to do."

     Throughout my tenure at Polaroid, I maintained copious **engineering notebooks**, kept the latest on my desk, took contemporaneous notes of meetings or phone calls, sketched design ideas with written notes, entered most of my calculations, and taped or stapled sample photographs and small blueprints into it. With authorization, I kept these notebooks when I resigned, but they were ultimately subpoenaed for the lawsuit against Kodak, and were never seen again.

     Over the course of the next two years or so there were several distinct camera development projects. Three were nicknamed Springer, the August camera, and the Christmas camera. None of these came to full fruition. Perhaps the greatest challenge for all these cameras was a **practical viewfinder and rangefinder** combination because of all the odd angles of the camera.

VIEWFINDERS and RANGEFINDERS – By this time my assistant Phil had fully joined the challenge of developing a good viewing system. Many configurations were hypothesized. Most anticipated an eye window either at the very back (as in Fig. 3 of US Pat. No. 3,683,770), requiring the camera to be held in a cantilevered position against the bridge of the nose; or near the roof-peak (as in Fig. 1 of US Pat. No. 3,610,123 also on back cover upper right), allowing the cheek to rest on the back roof surface, but requiring that the head be tilted forward and the eye rolled upward. Both of these positions were quite remote from the lens with the need of an optical or mechanical linkage from front to back for viewing, ranging, and parallax-correction. For rangefinder links, flexible shafts, levers, and hydraulic lines with tiny bellows at each end were proposed and modeled. All told, there must have been two dozen or more proposed solutions for framing and ranging, few of which were promising.

One of the rangefinders that Land encouraged was a stereo version that utilized both eyes for determining distance. In the end, I designed two versions, which were installed and tested on a 100-style camera for ease of demonstration (see US Pat. Nos. 3,622,242; 3,680,946; and 3,610,128, illustrated atop a ColorPack Land Camera). The first of these adjusted the convergence of a virtual target formed by an arrow-shaped reticle, and projected that target in such a manner that it would appear to move toward and away from the camera with focus adjustment. The rangefinding was accomplished by the photographer adjusting the camera until the projected arrow appeared to be at the same distance as the subject. It was quite accurate, but not entirely satisfactory to me. Therefore, I built another version that not only adjusted the convergence but also the accommodation (focus) of the target in synchrony with the convergence. Not only did the arrow appear to move toward and away from the camera, but it was always in the correct focus for the corresponding distance. This version was gangbusters. I even asked an engineer with one glass eye to try it. At first he laughed at me and refused to even look through it, but ended up being quite consistent with adjusting the range— presumably by comparing the focus of the arrow with the focus of the subject. Dr. Land reported to me that Edward Purcell, the Nobel Prizewinner from Harvard used this rangefinder and commented, "…best from Polaroid in years." As successful as the stereo rangefinder was technically, it never made it into the SX-70 Land Camera or any other Polaroid Land Camera.

The **first fully developed camera** design was styled with the help of Jim Connor of Henry Dreyfuss Associates, the industrial design firm in NYC that had styled all telephones of that time, many John Deere machines, Honeywell's round thermostat, *etc.* This camera folded down into a U-shaped tray running the length of the folded shape (see US Pat. No. 3,610,123 and back cover upper right). It was opened by lifting the viewfinder/rangefinder "bar" at the back end, thereby raising a pair of pivoted links that in turn lifted the other components into place. The finder bar had a telescoping section in back that I devised in a manner that could not jam—famous last words, right? But it did work. (See Figs. 3, 4, and 5 of said Pat. 3,610,123) This was the first working SX-70 camera to approach a realistic level of utility. It had all the basic pieces needed for a product, and some soft-tooling was being made for prototype manufacture.

The bar-type viewfinder/rangefinder had a distinct disadvantage with respect to parallax and accurate ranging when used with a lens that focused so close. Also, Land was never particularly happy with that model.

Therefore, he invented a single lens reflex viewer with an eyepiece at the camera's peak for looking down at a focusing screen just above the film plane. This concept required much greater complexity of both the mechanism with its multicycle shutter, flipping mirror, and folding viewer; and the optical system with its off-axis aspheric lens and mirror. The design and development for this optical system was largely done by Baker and the Henry Street Optical Division. (One piece of this new configuration, the film-loading door, can be seen in US Pat. No. 3,643,565, which covers and claims my misslatching-proof switch design for ejecting the dark slide. It is also seen in the lower left figure on the front cover.)

The Polaroid SX-70 Land Camera as it came to market less than a year after I resigned from Polaroid. The eyepiece of its single lens reflex optical system is at the top-right.

During the period that these two working camera configurations were emerging, the ninth floor engineering team was growing enormously, roughly from 1969 onward, as electronic engineers arrived to work on the shutter and flash, factory engineers arrived to assure manufacturability, optical designers arrived to refine the increasingly complex optical system, film machinery was being built in the previously-vacant half of the ninth floor, and some of us, including Wareham, Phil, and me, took on a technical overview role and began thinking about the next generation of camera.

I did not find this role particularly satisfying, was anxious to rejoin the center of product activity, was unsuccessful at finding it within Polaroid, and eventually decided to find it elsewhere.

• • •

So, **whatever happened to Polaroid?** When I joined the company, it had about 2,700 employees, enjoyed a splendid reputation, was recognized world-wide, had sales that grew handsomely every year, had stock that doubled about every two years (New York Stock Exchange, ticker PRD), and was considered one of the most benevolent companies to work for. Whenever I mentioned either "Polaroid" or "Dr. Land," the reaction was always recognition and often with a bit of awe that I was lucky enough to work at such a wonderful company. When I left Polaroid 6.3 years later, it had grown to about 17,000 employees, had about a billion dollars in cash despite the billion *each* it was spending on the SX-70 camera and its integral film system, and enjoyed all the same sales and benevolent characteristics except that the stock price was largely stalled. Whenever I told someone, either in or outside the company that I was leaving, the reaction was disbelief. "But why? No one ever *leaves* Polaroid. Why would you do that?" And, I sometimes wondered too.

After leaving, I didn't follow the company closely but tried to generally keep up with its fortunes, or lack thereof. The first momentous event, though of no surprise to me, was that the Polavision™ instant movie film system was a complete dud—Land's first market failure in decades. Surely, they could have read the early tea leaves about competition from Betamax and VCRs. Polaroid did, however, win their patent infringement case against Kodak and added nearly a billion dollars to their coffers (judgment was $873 million, final payment $925). The next spectacular event was that the younger officers via the Board of Directors tossed Land out of the company. New blood and plenty of cash—surely it will go somewhere now. But it didn't, and over the years I heard one story after another about its deepening troubles.

Of course there are many pieces to their puzzle of ultimate demise, but my view of the most basic underlying cause is this: Land ran a paternalistic operation in which he was the central source of almost every decision and product. Smart underlings were either of the yes-man type or they didn't stay at Polaroid for long. Bill McCune, the first president after Land departed, had been with Polaroid as a vice president for decades, was certainly a smart and capable person, but was surely one of Land's yes-men. When the chips were down, he really didn't know what to do. I never knew Mac Booth, but he had been with the company for years, too, and may have been of similar cloth. The old way of the 1940s, '50s, and '60s was lost to Polaroid, and the new blood just could not find a new way using the old ideas and methods learned at the *pater's* knees. It was a sad story that such a great American corporation could fail so completely.

• • •

And **what about that SX-70** that I designed—or at least got started? It turned out to be Polaroid's ultimate, not its penultimate camera, as Land once wanted to name it. Oh, sure, there were all those variants of the rigid Pronto and the later Spectra, but they were based on the same form of folded optical path and film relationship that Land and I devised over the phone that summer day in 1967. Bonanos, the author of *Instant*, told me that the number of SX-70 cameras produced was still unknown to him, but the total number including its successors that used the integral film was between 100 and 300 million cameras. Makes me simultaneously proud and melancholy.

OUTSIDE POLAROID and ECA
The Apartment House, Cambridge, Massachusetts
September 1968 to August 1974 – Ages 28 to 34

In June 1968, I got a letter giving notice that many buildings in my neighborhood were being taken by eminent domain for a new elementary school house. We were given four months to clear out. (Years later my daughter attended that new school, and I figured that my former second floor bedroom had been roughly where one of the new basketball hoops was located.)

I had already begun thinking that investing in an apartment house would be a good idea. This was the perfect catalyst to get me moving. In July, I found the ideal building. It was a two-family Victorian built around 1875 that had been converted to four apartments in the 1920s. The Inman Square area was unfamiliar to me, but it appeared to be among the nicer of Cambridge's typically run-down neighborhoods. The building needed work, but not so much that it would overwhelm me. It was priced at $27,000. I offered $24,000 but was ultimately conned into raising my offer to the asking price and never regretted it.

This was a big hulk of a house with little grace. It was 42 feet across the front, extended back 66 feet, and had a hip roof, which in my view makes any house look a bit frumpy. It was symmetrical with a three-window bay on each side to break the monotony of the front façade. Each side had an elaborate double-door entry. Every apartment had about 1,000 square feet divided into one bath and five rooms, four of them generous in the 12 by 14 foot range.

The very first thing I did even before "passing papers" was to research the issue of house numbering. For some curious reason, the two entrance doorways were numbered 95½ and 97, respectively. The house to the left was 95, the one to the right was 99. The ½ was confusing as it seemed to link to the neighbor's house and also had the stigma of a rear entrance or a basement location. It would be less confusing as 97 and 97½, but I didn't want to keep the stigma. No one seemed to be in charge of such things, so I just told the Post Office it was changing to 97 and 97A. I hand made some nice looking 5-inch high brass characters in the font Eurostyle Bold Extended (also known as Microgramma) to adhere to the glass transom above the doors where they would be back-lit at night. The numerals remain there today, almost half a century later, though no longer polished. (I've continued to use that font on most of the houses I've owned since that time, but not always in bold.)

The most-urgent issue was installing central heat. Every apartment depended on a space heater integral to the kitchen stove. One of those heaters was fired by kerosene that was supplied by a can of fuel inserted similarly to the inverted water bottles on office water coolers. It was a wonder that the house hadn't already burned down. Moreover, neither I nor another new tenant had any space heater, and I didn't want to purchase two just for a temporary measure. Therefore, I contracted Cambridge Gas to install hydronic heat in every apartment from a central boiler. And this had to be done before the cold weather set in.

Another urgent matter was to free-up one of the apartments to move into myself. I had to vacate Hayes Street by October 10, and the closing was scheduled for September 19. Tenants didn't have *rights* in those days, but it still didn't give me much time to evict one of them. Somehow I learned that one of the tenants was moving out at the end of September. Whew! But that wasn't the apartment that I preferred, so I came up with a plan.

The plan was to move temporarily into the available unit, completely clean, repair, and redecorate the available apartment so that the other family could switch apartments with me, giving me the unit that I preferred—second floor, sunny side. Ultimately it took three months to get that apartment into shape for the other tenant. Meanwhile I was living in the mess of wallpaper scraps, joint cement, sawdust, paint flakes, drop cloths, new floor tiles, plumbers, electricians, rearranged kitchen appliances, *etc.* But in early January, we traded apartments, and I could begin to think about settling.

My friend Doug from Polaroid had been largely responsible for getting me interested in owning an apartment house. He had owned several over the years. Once when he was talking about his experiences, he mentioned that it was unbelievable what these old Cambridge houses might have in them and proceeded to relate the example of a kitchen sink drain that had been repaired so many times that it was encased in multiple types and vintages of electrical tape. I assumed that story to be apocryphal, but sure enough one of the first tenant complaints was about a leaky drain that turned out to mirror the taped-up drain of Doug's story.

Speaking of drains, a couple of months after purchasing the building, a second tenant moved out leaving me with a decent apartment except for the kitchen sink. The sink was a cast iron enameled basin set onto a framework of 2x4s with a thumbtacked curtain hiding the frame. The water was supplied through a pair of rusting galvanized pipes running horizontally across the surface of the wall with a garden-hose type of faucet installed in each pipe; thus one faucet was a little higher than the other. When I departed from the Hayes Street apartment, the city had told me I could take *anything*, including doors, appliances, sinks, and toilets. I had taken two of the kitchen sinks, one for this very apartment, and now was faced with installing it. I traded out the sinks, and a moonlighting plumber came that Saturday to install new supply pipes. Now, the used but much better kitchen sink was fully installed except for the drain pipe.

By this time, I had already advertised on a bulletin board at Harvard. The first call was from a girl with two roommates. Just before she arrived, I rushed upstairs, put a bucket under the sink drain, and pushed the drain basket securely in place—double protection. The girl was very nice and said almost nothing as she slowly looked over the entire apartment with her arms akimbo until she finished up in the kitchen. For the first time she unfolded her arms and turned on the water for just an instant. I was so relieved that I had pushed the drain basket down, but she looked at the water a few seconds then reached in and pulled up the basket. Of course she heard the water splashing into the bucket, opened the cabinet to see the bucket, and gave me a quizzical look. I knew that nothing but humor could save this, so I said, "Oh, no problem, you can empty the bucket in the bathroom sink just around the corner." She couldn't help but laugh. And she rented the apartment. I made sure the job was finished before they moved in.

I learned a lot about caring for and restoring old buildings. For years I kept a running list of projects throughout the building. Sometime later I used the list to estimate that I had put well over 2,000 hours of my own time into that place, mostly during the first two years that I lived there. One apartment took about three months of my evening and weekend time to repair walls and floors, have electrical and water upgraded, improve the kitchen, and paint throughout. Another apartment got by for a few years with little other than that already-described kitchen sink replacement. But the following tenant wanted better so he, his cousin who was another tenant of mine, and I worked to install a completely new bathroom and mostly new kitchen. In a third apartment, I installed a modern but used toilet in place of the failing overhead-tank-with-chain type. The subsequent tenants in that unit also offered to help with some of the work, and improvements for them included both a new bathroom and a partial upgrade of the kitchen. I also hired that tenant to rebuild two of the four back porches.

• • •

In my own apartment, my first project was to move the bathroom three feet so that the hallway from the kitchen, past the bath, and into the bedroom was rerouted from the inside wall to the outside wall. This resulted in having a closed corner in the kitchen so that a continuous line of cabinets and counter space could be installed. Previously, as in all the apartments, there was a door in every corner. It also resulted in a blank wall in the bedroom so that a room-width closet with bifold doors could be built instead of the undersized boxy bump in one corner. Moving the bath required having a lot of 4-inch cast iron soil pipe drainage and revent lines removed and replaced with 3-inch copper pipe. It also required several days of climbing through the bathtub to get into and out of bed. To make the job still more challenging, I chose to install $^5/_8$ sheetrock for sound-blocking and to create an antique-style bathroom with restored Crapper-style toilet with pull chain, a footed tub sitting in the equivalent of a shower stall for easy showering, a marble sink counter with polished brass fittings found at Dwain's junk yard in Quincy, terra cotta-colored quarry tile, and a restored, heavy, sound-blocking door with wooden knob also found at Dwain's.

The 11 by 13 foot bedroom was finished next, and needed patching of walls and floor and new sheet rock on the ceiling and two walls. I also built the new closet. The floor was carpeted by pros. The 13 by 16 foot kitchen remained in a temporary state of disorder while outside summer projects and other activities took precedence.

One of those activities was getting married. (For more, see *My Family* in the *Adult Experiences, Tales and Anecdotes* chapter below.)

But in the fall, the kitchen project resumed with new sheetrock on the ceiling, removal of the plaster from and varnishing of the artful brick firewall, having the plumbing and electrical circuits upgraded, assembling and installing about ten do-them-yourself fir cabinets on the badly-sloping floor, making the L-shaped Formica countertop from scratch and without proper tools, and installing vinyl tiles over new floor underlayment. The pantry also got new shelving including a section for wine bottles with a diamond pattern of 30-degree sloping shelves. I also built a dining table of old wood from Dwain's— probably yellow pine considering its enormous weight and hardness.

After remodeling the rear half of the apartment, I undertook the front half.

REMODELING THE FRONT ROOMS – Originally, the front door at the top of the front hall steps opened into a narrow, dark, and useless hallway that led rearward into the kitchen or frontward into a small room that I had adapted as an office and light-duty workbench area. An awkwardly-offset door in the wall of the hallway virtually isolated the entry from what should logically be a living room. Thus the living room was nearly inaccessible and made uglier by a bump in one wall that enclosed a then-unused chimney. My project was to remove that wall and the adjacent closet-wall to make a much larger and open space for a nicely-located living room. Thus the size would increase from 13 by 15 to nearly 17 feet square.

It worked. The chimney ended up passing through the room, but well off-center, and I "decorated" it by varnishing it with glossy polyurethane and installing a clock through the chimney so the works were at the back and only hands showed toward the main part of the room.

Of course the floor needed repairing in the places where the walls were removed, and the wall next to the entry door needed some work, which was when I discovered that unlike the rest of the horsehair-plastered house, this wall was made from sheetrock, and the date on the back of the sheetrock was from the 1920s suggesting that the two-family building had been converted to a four-family at that time. Thus, our apartment comprised the original three upstairs bedrooms at the front and servants' quarters at the rear. The ceiling was a mess. The original calcimined ceilings had been painted over which caused disastrous peeling, the removed walls left the remains of the stud-cap which I left in place for structural reasons, and there were slight differences in the several levels. The solution: install a suspended ceiling but of unusual construction. I made panels about 3 by 5 feet from lattice stock stained walnut and gapped at about $1/4$ inch. Also, the conventional main-tees and wall-ells were covered with identical pieces of lattice. Thus the entire ceiling was of a similar texture of stained wood with little evidence that it was in separate panels. The plaster and patches above were all painted flat black to obscure them. Wires for hanging light fixtures and speakers wandered unnoticed above. After painting woodwork and walls, the floor was carpeted.

The front-most room faced southeast so I built a planter for its bay window. The planter was elevated to just below the window sills and its 4-inch deep tray-like top was waterproofed with tar so that it could be filled with potting soil and watered. Eventually, the window was covered with many varieties of plants including several orchids. The knotty pine floor in that room was successfully sanded and varnished and the walls papered. A designer in our Polaroid office built a bookcase to my specs, which was permanently installed in one corner and accommodated the stereo system.

Not long after the major interior work of that apartment was completed, our daughter was born. I was glad that she wouldn't have to endure growing up in plaster dust and learn to crawl over rusty nails and splinters. The front room with the planter became hers. And speaking of crawling, in her early stages of learning to walk, she once fell when she got to the single five-inch step from living room to kitchen. Henceforth, she would toddle to the doorway, get down on all fours, back down the step, and stand up again. Within a month or two of her learning that trick, the weather got warm enough to keep the back door to the porch open much of the time. There was a much higher step here. But by this time she was more confident and learned to negotiate that ten-inch step while still standing. Despite this newly-learned skill, she continued to negotiate the smaller, inside step the *old* way in a crawling position.

• • •

This house was the source of many interesting and challenging encounters and projects, a few of which I'll mention here.

After the living room was finished, I kept my reclining chair next to that newly-exposed chimney. I liked looking at the utility-grade bricks and the rough, unfinished mortar work used for that long-hidden chimney. One day, I had a sudden insight—the bricklayer had been left-handed! Suddenly I knew something personal about one of the builders of that house a hundred years earlier. I was so delighted! How could I know? Well, I had spent many hours in my youth watching the construction of houses, and one of the crafts was bricklaying, so I knew how they buttered mortar to the end of a brick, how they removed oozing mortar from the horizontal joints, and how they sometimes rough-pointed the joints. Since these mortar joints had never been neatly pointed, the rough mortar joints revealed the original operations and motions of the trowel. This trowel had been held in the mason's *left hand*.

One evening as dusk was settling, I dug a 10 or 12-inch hole in the back yard to plant a bush. With the last shovel of dirt, a golf ball-sized clod rolled off the shovel, into the hole, and disappeared into a dark spot in the bottom of the hole. I stared in disbelief, decided that the dark spot was indeed a cavity, inserted the shovel handle into the dark spot, and freely plunged it for its full length into the ground. I was so disturbed by this that I quickly planted the bush and turned in for the evening. A year or two later my neighbor told me that there was a car buried in the back yard. I never did revisit that area of the yard and always wondered if my rose bush was planted over a rusting four-foot deep cavity in the ground.

Each apartment had its own water heater in the basement adjacent to another apartment's heater. One time when a heater failed, I made a quick fix by interconnecting the good heater to the other tenant's pipes until the plumber could replace the heater a few days later. The trick worked so seamlessly that I installed a permanent interconnection in which the "valves" were made from unions in which I could insert or remove a disc of copper to stop or allow the flow. Henceforth, I could effect a fix for the failed heater in about five minutes in a manner that was completely obscure to the tenants. It also allowed the new heater to be installed without paying emergency rates.

Late one night after I had moved to Belmont, I was called to clear a clogged drain. The clog was just below a 4-way connector that had such a small amount of sweep that a snake inserted from the kitchen went up into the bathtub and vice-versa. No matter how I shaped the tip of the snake, it would not make the turn and go downstream to clear the clog. It needed to be inserted from the vent side, but that was three stories up on the roof. It was late...I went home to sleep...and woke with a solution. I would drill a hole in the side of the vent pipe for inserting the snake and just patch the hole in the black, water-free pipe with a few wraps of black electrical tape when finished. When I went back and surveyed the best spot for drilling the hole, I saw that my finger had landed on a ring of black electrical tape. I wasn't the first with that brilliant idea!

The tenant on the first floor of the 97A-side once called to report that there was an animal inside the outer wall of their living room and that it was chewing a hole into their room. He said the hole was very large, about three feet, in the middle of the wall, that he could not see enough of the animal to identify it, and that he was afraid it would get into the apartment. I agreed to take a look after work. My vision of his report was of a 3-foot diameter hole in the plaster about shoulder height to one side of the window. There must be two or three studs visible. The pockets are only 4-inches deep. How can this be? What kind of animal could make such a hole and still be partially hidden? After work, I got my toolbox and drove into Cambridge. As I approached the front steps, the tenant from the second floor apartment on the 97-side emerged and I stopped to chat and ask about his wife and son. He reported that "Damien is very sad—his gerbil escaped a few days ago and disappeared." A little bell went off in my head, so I invited him to go with me to investigate this animal hidden behind a 3-foot hole. Sure enough: the gerbil had escaped from Damien's attic bedroom, found its way through the dividing wall to the other side, crossed the attic floor to the far wall, and fallen two stories inside the "balloon-constructed" wall to the level of the first floor. The "three feet" was the length of a small gap between baseboard and floor, the "middle of the wall" meant front-to-back, not half way up the wall, and the gerbil had gnawed only a short section of the opening just large enough that it could, maybe, squeeze out. Damien was summoned with his cage and some food. The gerbil eventually came out. (And I never got around to patching that little gap.)

• • •

In August 1974, the three of us moved away from that apartment house to the Greek revival house in Belmont that we had purchased a year earlier. I continued to own the apartment building and maintain it until 1987 when it became too exasperating to deal with rent control and increasingly unknown and short-tempered tenants. Too bad: it must be worth a small fortune today.

ELECTRONICS CORPORATION of AMERICA
March 1971 to September 1975 – Ages 31 to 35

He was turning thirty. He was feeling adrift. Was he being useful? Had he accomplished anything? He was doubtful of the answers. The true answers seemed to be *no*. He had to do something about that.

So, he set about to change things and to get going with his life. But, where should he go? Was his original career choice the right one? What did he *really* like to do? What was he *really* good at doing? What did he do best? What made him the happiest?

Well, the *he* was *me* and I was having my first life crisis...at thirty. The reality was much different from the internalized impression of myself. My career choice *was* right. I had been an engineer since early childhood when I began reworking cardboard boxes into houses, and I had always preferred and enjoyed the physics in my life more than any of the other aspects. I remember once being told by a much older, philosophical acquaintance that I would eventually learn that "ideas are more interesting than things," but the reality for me had always been and would continue that "things" were more interesting.

In any event, at the beginning of my fourth decade, I explored myself and my life's work more than at any other time. I was dissatisfied with much of what I found. I observed—an observation that remained true for me always—that my professional work, my "day job," was of major importance to my happiness and success. I tried to move into more adventuresome areas within my then-current employment at Polaroid but couldn't make it work. I talked cautiously to my own boss, and he counseled patience. I explored the possibility of going into real estate development because of the satisfaction recently gained through the ownership, landlordship, and remodeling of the four-unit apartment house bought a few years earlier. That looked doubtful—at least through the channels explored. I considered, but never seriously, whether I should go back to school in preparation for a completely different career path.

What did happen is one part of the story of my life. It was both glorious and tragic.

I went to Electronics Corporation of America on March 4, 1971. I remember that date as being noteworthy—it was very hard to "march forth" from Polaroid. Few people ever did it. Polaroid was such a benevolent place, with such nice surroundings and truly stimulating people to work with that it was rare for a person to leave voluntarily.

The day of my interview was very cold. I felt particularly cold as I tried to be casual while leaving Polaroid in early afternoon. At ECA, they kept me waiting in the lobby for quite a while. I even asked the receptionist if they had forgotten me. Finally a secretary came to escort me to the engineering office. The third floor was nothing like the beautiful spacious, two-story reception hall. It was a huge dingy top floor of the former Filene's cold storage and warehouse building. The floor was concrete covered with worn and chipped layers of paint; the ceiling was the concrete structure of the roof with exposed pipes in all directions; and the windows were those large metal industrial tilting sashes with hundreds of mismatched frosted glass panes and a few clear, but grimy, panes along the bottom two rows. The offices were cubicles partitioned with grimy six-foot high metal and frosted glass panels.

The central portion of this floor was occupied by rows of technician's benches at the near end and by drafting boards at the far end, which ended in a point where Memorial Drive met Main Street at the foot of Longfellow Bridge. The view across the river to Boston was gorgeous—if you could find a clear pane to look through. At least the cubicles were graced with vinyl tile on the floor.

Most startling upon entering the engineering department was an impressive, dapper-looking older man at work amongst the technician's benches. He was completely bald with a close-cropped full beard; he wore a white shirt, bow tie, vest, dress trousers, and had his sleeves rolled up two turns. He was handling a long piece of flexible metal tubing, and as I passed, he looked up at me, smiled, and nodded. I remember thinking: "with an engineer who dresses that way, the company can't be all bad."

I met with Phil Cade, Engineering Vice President, who was friendly, cordial, and asked some insightful questions but was otherwise somewhat flat. I didn't meet with any of the other engineers or managers, and Mr. Cade excused himself at one point and left the office for a while. By this time I had decided that this was no place to work. They were not treating me with much respect, the place was a dung heap, and what few projects he showed me were primitive in comparison with anything I had worked on since a summer job many years earlier. But when Cade finally returned, he chatted a bit more then asked if I could stay long enough to meet the company president, Dr. Metcalf.[9] By this time it was too late to return to my Polaroid office, so I agreed.

The executive floor was much nicer with carpeting and wood furniture. The real surprise was that the president turned out to be that dapper "engineer" that I had seen upon entering. Metcalf was a charming and inquisitive man. Among other questions, he asked, "Are you ambitious?" And after I replied affirmatively, he asked, "Ambitious to do what?" I answered, "To accomplish things." That must have been a good answer for he hired me on the spot and for my asking price, which was 25% above my Polaroid salary.

I must have been naïve: for when I met him as president, I just assumed that he was a hands-on sort of president, somewhat like Land, and had happened to be working with a technician when I arrived. Much later, I realized that all this had been orchestrated: the difficulty in establishing an interview date, the long wait in the lobby, and the absence of Cade during my

---

9 Arthur G. B. Metcalf, founded ECA in 1937, was 62 when I arrived, had received engineering degrees at MIT and Harvard, and taught aeronautical engineering at BU for a short stint before forming ECA. At barely five feet, he was short in stature, but very large in presence, and ruled his company with a quick and furious temper, sometimes followed by a firing. He rose to lieutenant colonel in the Army, thus using the honorific "Colonel Metcalf" for years before receiving an honorary doctorate whereupon he switched to "Doctor Metcalf."

interview. When Cade got my application, he must have brought it to Metcalf's attention. Dr. Metcalf must have thought this too good to be true. Though I never knew for sure, Metcalf was undoubtedly jealous of the success of Polaroid. The two companies were nearly the same age, had prospered within blocks of each other, but ECA was only 7% the size of Polaroid and had no public persona. Surely he must have thought that snatching me away from Land and Polaroid would be a delicious coup.

I took the job. My announcement at Polaroid was met with disbelief. Land asked if I wanted him to persuade me to stay. I said, "No."

Although my boss, Dick Wareham, was devastated, he also seemed to be the only person around who understood my ambitions to take on a new challenge. I sensed that he empathized with my wanderlust and admired my daring. Ultimately, Dr. Land even decreed that I was to take a one-year leave of absence so that I could easily return to Polaroid if the new job did not work out.

Cade treated me well. Each day that first week, he took the time to fill me in on something new: my first project, introductions to my engineering colleagues, the company's organization, an overview of ECA's Photoswitch™ and Fireye™ products, and a rundown of all those people on the distribution list of the announcement that I had joined ECA as its **Chief Mechanical Engineer**.

My office looked out over a couple of streets and the Charles River beyond. How ironic: one of those streets would later be named Edwin H. Land Boulevard.

I was immediately overwhelmed with projects to do. At the end of the second week I made a list summarizing my project assignments thus far. It had 19 line items. Admittedly, some of those were very brief, but there were five or six projects with real challenge, such as the new smoke detector product and the conversion from a die cast 42-series housing to a plastic housing. None of these matched the SX-70 in scope or challenge, but all of them included the enticement of being small enough that I might actually bring something to market quickly, efficiently, and with my "stamp" on it.

The people there were as nice as those at Polaroid. The designers, drafters, and documenters were helpful and eager enough to accomplish things that they provided quality support and results. The model shop could not compare with those at Polaroid, but they were adequate for the projects being undertaken and the machinists were entirely competent.

I had been working there for about a month when Dr. Metcalf (I always addressed him as "Doctor Metcalf") showed up one day for an update on the **smoke detector**. That device was modeled on their most common line of "42-RL" photoelectric switches, which used light bulbs as their signal source and were installed in a 3 inch diameter housing about 8 inches high. The smoke detector was to have an internal photocell to monitor the reference brightness of the bulb, and the circuit was designed to respond to a 5% reduction of the reflected light. My predecessor had designed some awkward hardware, and the electronic circuitry had already been designed. This project just needed the hardware tuned up, made producible, and documented as a product.

I was taken by surprise when Metcalf, Cade, and two managers descended on my office and crowded in. I had been comfortable in Land's presence and accustomed to his surprise visits, but this was the first instance of such with Metcalf. I stood near my drafting board, attentive, but mostly silent. Metcalf asked a lot of questions and made a lot of pronouncements, most of which were answered by Cade and a few by me. It was clear that he was not happy with progress. As he got up to leave, I attempted some levity and blurted out, "We may not be fast, but we sure don't do good work!"

From the corner of my eye, I could see everyone else in the office freeze with expressions of terror. Metcalf paused with me in his eye for a couple of seconds. Then a twinkle came into his eyes, followed by a smile and the single word, "Exactly!" He departed.

Later, Cade came into my office and offered me the advice never to do that again. He even seemed surprised that I still had a job.

As mentioned before, I began working at ECA in March. The offices were grimy and noisy, but survivable. But when one of the first really hot days of spring arrived, I was in for a huge surprise. The offices had no air conditioning, and no insulation to separate us from the black tar on the roof. The only solution to cool the area was to open the windows. But, while the breezes from the Charles River felt good, they blew the settled dust everywhere and caused the Venetian blinds to rattle incessantly. Moreover, the open windows admitted the noises of trucks downshifting to make the hill over the Longfellow Bridge and the roar of subway trains emerging from the tunnel and echoing off the walls of the recessed tracks. I remember after a few hours of this torture, pausing to hang my head on my desk and lamenting, "What have I done by leaving Polaroid?"

One of my fairly early assignments was to help an elderly engineer with his prototype of the **extended scanner**, a water-cooled two-channel ultraviolet detector for use deep within the buckets of a large utility boiler. Soon after I began this partnership, he became sick with cancer, so the scanner became my project. Someone else handled the electronics, but I had all the rest. Soon thereafter, I planned a full temperature test-run by Jim, a technician in the furnace lab. Jim was a great experimentalist and quickly constructed a test chamber that heated to 1,100°F. The water-cooling circuit was a major failure, and the UV tubes were badly damaged. Jim had a great suggestion for fixing it, which I immediately began designing into the product. While I was working on the design, Metcalf dropped by to ask how things were going with the scanner. I described not only the failure, but how we planned to improve it. He complimented me on such insight, but I was quick to protest that this was Jim's great idea. The next day, Jim told me of Metcalf's visit to his dungeon of a lab to thank him for his valuable contribution to the project. Jim became a loyal supporter of mine, and I also sensed that Metcalf developed a greater respect for my integrity, too. That was a good lesson for me.

There was no doubt Metcalf was a temperamental, egocentric man, demanding much and expecting everything. Many **stories of his antics** and temper were noted. Three were especially memorable to me.

ANTICS – The extended scanner was a favorite of his, and one day he came into my office to see how it was progressing. I reported that its design was complete and was about to be released for a 30-unit pilot run. He was pleased to hear that and pored over my drawings, asking several questions. At one point he wanted the opinion of Andre TerMeulen, VP of Marketing on the potential sales forecast. He turned to my phone and dialed the operator. Of course the operator was not expecting Metcalf's calls from *my* phone, so there was no immediate response as he was accustomed to. After a few seconds, he barked to me, "What's Andre's number?" I stumbled to the phone booklet and looked it up: "512" (I can still remember it 40 years later). He dialed, and Andre's secretary reported that he was in Mr. Brennan's office, so he put the phone down and dialed the operator again. I knew trouble was brewing, so I quickly looked up Brennan's number in the booklet still in my hand. Again, a few seconds waiting for the operator, then he took my proffered number and dialed Brennan's office. The report was that he had returned to his own office. By this time, Metcalf was getting pretty red in the face. "What's Andre's number?" he demanded of me, and again he dialed, impatiently this time. His secretary now reported, "Andre just left for the day." At this point he exploded, stood up, and yelled: "But... but... but... a minute ago you said he was in Brennan's office" while sputtering spittle all around my office. Even before Andre was found and redirected to my office, Metcalf was crossly asking me more questions, one of which was why the water-cooled parts were made of aluminum instead of stainless steel? He didn't buy my answer that aluminum had a much higher thermal conductivity than stainless and demanded that "our product should be made of stainless steel." All those carefully-selected tubing sizes available in aluminum had to be reselected from the dissimilar choices of stainless, and all those drawings about to be released had to be revised or redrawn over the next month. Amazingly, the final overall thermal performance of the stainless prototype was equal to that of aluminum sample. It speaks volumes about some people's instincts despite their ignorance.

Once when Metcalf was away for a couple of days, a sewer pipe broke in the basement of the building. A septic contractor was hired to pump out the mess and repair the drain pipe. It took several days to complete, and was still in progress when Metcalf returned to the office, arriving as usual in his Rolls Royce and wearing one of his natty three-piece suits and typical derby hat. He spotted this filthy septic truck backed up on the pristine front lawn. As he passed the receptionist, he ordered that she send Carl up to his office. (Carl was Metcalf's fix-it man.) Carl arrived and Metcalf demanded, "What's that truck doing on *my* front lawn?" Carl explained the problem and that the front window was the only possible access to the pool of sewage accumulating in the basement. At which point Metcalf's face got yet redder, and he began to sputter and blurted, "Well... Well... Well... At least get the truck *straight* with the building!"

In my last year at ECA when I was Manager of Manufacture and had an office next door, I often found myself dragged into meetings with him. On one particular occasion, about four of us were sitting around his coffee table facing Metcalf at his traditional seat in the center of his black leather sofa. Over the course of the meeting he got so worked-up over some issue that first he got a headache and called for his secretary to bring him an ice pack, then he developed a nosebleed and stuffed a Kleenex into one nostril, all the while yelling at us and pushing back on the sofa with enough strength that the front legs rose from the floor several inches. Here were four adult men gathered around with serious looks on their faces intently following the red-faced ranting of this wild man rising up and down on his sofa with an icepack on his head and a Kleenex flopping from the middle of his face. It was all I could do to keep a straight face.

But Metcalf also had his good moments. Example: In early 1972, Andre, the VP of Marketing came into my office one day and said, "Hey, look at this." and pulled out a Hewlett Packard **HP-35 calculator** from his pocket. Pocket calculators were completely unknown at that time, and I was astounded. After a few moments looking over the keyboard and trying a simple calculation, I realized that the options, keystrokes, and reverse Polish notation were identical to those on a console-computer calculator I had occasionally used at Polaroid. Some years earlier, I had derived the equation for calculating the radius of a lens surface based on diameter and sagitta (depth) of the curve, and had memorized the keystrokes including memory buttons and square root buttons. I proceeded to do a sample calculation for Andre. He was so impressed that I had figured out how to use some of the advanced functions that he reported my expertise to Metcalf. A few weeks later, Metcalf called me to his office along with two other people, both VPs of engineering, and gave us each an HP-35. Mine was serial number 1050, from the initial production run of 2,000 units and sold for $395. I retired my slide rules pretty quickly after that.

Whether it was because of project success or some other reason, I was invited to one of the day-long summer sailing expeditions on his 60-foot 2-masted sailboat that first summer and to the annual Christmas party the following winter. For the most part, only executive level or long-termers went to these events, so I felt honored to be included.

After I had been at ECA for a little over a year, Metcalf hired a new vice president to be the senior engineering manager—even over two vice presidents who had been working at ECA their entire careers. Chris Peek had spent most of his career at Sylvania developing many of their innovative lighting products. Although he was an irascible sort, we developed great respect for one another, and he gave me lots of support. He came back into my life several years later when he was instrumental in my joining GTE Laboratories.

THE FACTORY – As at most jobs, I liked looking at the **factory operations**.
ECA's factories were relatively simple and relied on manual operations for much of the productivity. The first floor included a fabrication area for stamping and forming sheet metal, turning round parts, tapping holes, molding plastic parts, and nibbling parts from large sheets on their new numerical-controlled Wiedeman stamping machine. There was a stockroom with hundreds of stamping dies, but only six or eight presses in which to mount them.

Scattered about that old building were other operations to watch such as a continuous electrostatic dip-painting process—always ECA grey; an automatic glass bulb-making machine for the UV detector cells; adjustable wire-measuring and stripping machines; and several stick-winding machines for making transformers that required continuous hand labor.

Upstairs, the assembly operations were done at assembly belts running at one to two minutes per station. Tooling was simple: rivet machines, air-powered screwdrivers, part-cradling fixtures, and two rarely-used resistor-insertion machines—hand-operated and only one resistor per setup. Most assembly runs were short. The longer assembly runs were done in Puerto Rico where the belts typically ran at 30 to 45 seconds per station. They were fast. I went to the Puerto Rico plant about a dozen times.

I worked on a slew of projects at ECA over the course of my four years in the engineering department, but four stand out from the rest because I had major engineering and management responsibility for getting them designed, released, and onto the factory floor. I have already mentioned the **extended scanner**, a project I took over in prototype stage from an engineer who died. I was able to complete the design, test it in a chamber glowing at 1,100°F, site test it in a power plant at Steamboat Springs, Colorado (while everyone else at the hotel was skiing), oversee the production drawings, and integrate the scanner head and cable with the electronic controls, water-cooling lines, and air-blast clearing of its sight lines. (See US Pat. No. 3,825,913 in which Fig 1, also shown on front cover lower right, is a copy of my hand-drawn perspective sketch.) This was early in my career there and although the first production run was only 30 units, it was very rewarding to see about $150,000 worth of inventory stack up. This project also earned me an invitation to the executive dining room where lunch was served to us at no cost: complete with special desserts and linen napkins.

At the time, all of ECA's relays were mechanical, but there were rumblings in the marketplace of **solid-state relays** coming soon. The company decided to be a leader, and a coworker designed the electronics for a double-pole relay using triacs, along with a lot of control and spike-suppression circuitry. I designed the hardware to match a line of plug-in mechanical relays, but with gray instead of clear covers. The biggest challenge was getting the heat out, but my heatsink design served nicely. We had all the parts, assembly fixtures, packing boxes, and labels ready for a trial production run of some 200 units. The only problem was cost. Peek had sold this project with hyperbole such as: "electronics is now so cheap with gates at a penny each…" When our relay design was ultimately priced near $60 with all its high Cambridge-based markups and overhead, Metcalf flew into a rage, "but you said pennies," and promptly cancelled the project with no option for reconsideration.

Perhaps the project with the most fun was a **flame-detector** that an electrical engineer and I did together. He had developed a novel technique for detecting flames accurately, and management decided to manufacture 100 of these devices within a few weeks for a valued customer's special project. Problem: there was no device—only a breadboard. The question brought to me was could I design a product robust enough for a boiler room environment including attachment to a sight-pipe, an optical system, a sizable circuit board, cable connections, *etc.*, quickly enough that 100 units could be manufactured within two weeks? Wow, I worked on the SX-70 for five years and it still wasn't near production. But, hey, what a challenge! Since I had been at ECA for nearly four years, I was familiar with almost every product and production machine they owned. I used this knowledge to ferret out a housing and most other hardware needed except for a special left- and right-bent chassis pair and, of course, the custom circuit board. Except for the bends, the two chassis were identical, and every shape and punched hole could be stamped using an on-hand tool in their numerical-controlled stamping machine. The photocell was captured between the pair without needing any additional support bracket. We didn't quite make the two-week schedule, but most of the 100 units were finished before three weeks had passed. That was a thrill.

But the most comprehensive project was the **world's first LED-based photoelectric switch**. It used the 42RL1 housing that I had designed in plastic my first year there. Originally, my part was minimal: I designed the optical spider that held the main lens, the infrared LED, and the photocell receiver, all in a manner that permitted adjustments in case the component specifications changed; and I designed the terminal panel to allow for the inclusion of a larger circuit board and a selection of plug-in timers. Periodically, I pitched in to help with the mold design or resolve some other peripheral issue.

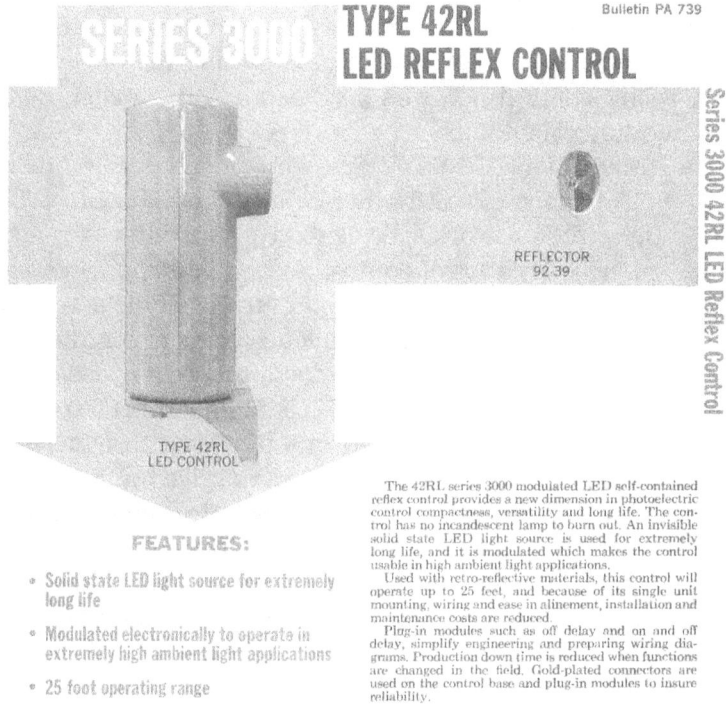

Upper portion of a cut-sheet for the 42RL1-Model 3000 for which I designed the plastic version of the formerly die cast housing, later designed the LED-specific optics, terminal board, and plug-in timer boards, and still later managed into pilot production in 1974.

Then in May 1974 Peek and I were flying together from Puerto Rico when he shared his dilemma with respect to the typical May layoff. His least distasteful layoff list included the engineer who was overseeing the pilot production run of 150 units of this new product, but he had no one to replace the man. I immediately volunteered. He was hesitant based on my lack of production experience, but I assured him that I was up to the job. He gave it to me, but it turned out that the project was in a hopeless mess: the two major molding tools with 14 week delivery times were not even ordered, no assembly fixtures or procedures planned, no test equipment designed or built, and most parts still not ordered. So much for "one month away…" I structured a schedule based on best efforts promised by others, rescheduled the pilot run for about four months hence (instead of the one month Peek had been promised), and followed up on progress almost daily.

Over that intense summer, the schedule slipped only about one week, and by mid-September, I was headed for Puerto Rico with the last of the parts and test equipment. I helped set up the line, and early one morning it was ready. About 40 women converged alongside the assembly belt, the forelady familiarized them with procedures with some help from me, and the belt started up. It was a 45-second belt, there were a few delays, but well before lunchtime there was a whole pallet full of product being wrapped up for shipment to the New Jersey warehouse. What a gratifying feeling!

Though not one of the outstanding projects, there was one with a most unusual and intriguing twist—**industrial espionage**. I was sent to a factory in New Jersey to learn how they attached leads to a ceramic substrate prior to attaching a solid-state circuit chip. This was certainly outside of my usual routine of designing gadgets from scratch. It also required powers that I had never known that I had. Incidentally, this was not a hostile activity—the factory had invited us to learn from them, but did not want to just give their design drawings to us. The crucial machine was about the size of a sewing machine, and I probably spent 20 minutes total looking at it, absorbing its details, and trying to remember dimensions as well. I couldn't just peer long enough to get the gist and think "that all makes sense." I had to concentrate on each piece, think about and remember its dimensions and how it was attached to or related to the rest of the machine. Even the drive motor and gearbox: I had to crane my neck to read the specification plate for speed and power. I didn't make any sketches or notes, but immediately upon my return found a drafting board hidden away in the back corner so I wouldn't be interrupted and spent about three days reproducing what I had seen. A few weeks later, parts came back from the shop, and my technician assembled a working "substrate staking" machine.

• • •

When I had been there for four years, Metcalf had a shake-up on the executive floor and his long-term executive vice president became *persona non grata*; he set up an office of three assistants: engineering, quality control, and manufacturing; and soon hired a new VP. I was invited to serve as **Manager of Manufacture and Special Assistant to the President**. Although I felt ready for a management role, it should have been in engineering, not such a high-visibility one outside of my forte. But I also knew that Metcalf would not take a rebuff lightly. I would have to accept.

The job was somewhat goofy. Three vice presidents reported to me although I wasn't one myself. I had no personnel administrative responsibility for any of them or their subordinates. I had already known and worked smoothly with them, and they responded to me well in this new position. I had barely gotten settled in the job when the new executive VP was hired. He had all three of us running around working on useless projects and demanding that we meet our promises to the day. For example, one he gave me was to determine what fraction of the Cambridge building was being used. I spent days wandering around looking at space, making measurements, calculating square footages, and determining which departments were responsible for these empty areas. The result was a table of impressive information, but when Metcalf heard of this he blared, "What kind of nonsense is this? Any idiot can walk around the building and see in a few minutes that *some* space is not used. What would you expect?" Needless to say, that VP didn't last very long either.

About three months into my new position, Metcalf suddenly got the impression that the Puerto Rico assembly plant was operating inefficiently and he assigned me to reorganize it, but he imposed the rigid guideline that no product was to be running on a production line for less than one week at a time.

> **TUNEUP** – Previously, products had been manufactured in batches equivalent to three-months of sales. This new guideline would result in some products having hundreds of years' worth of inventory. Only one of the 250 products made there could meet both guidelines. How could I tactfully resolve this crazy dilemma? To make it worse, Metcalf departed for the summer on his sailboat. On the occasions when he returned—once every week or two—he asked me for a progress report. When he heard of the results, he would throw in an alternative limitation and send me on my way. Each of these proved to be equally unworkable. After several cycles of these absurdities, he agreed to one of my suggestions: assemble *groups* of *similar* products for a week at a time. Since I knew the hardware much better than the scheduler, I set up the groupings. I also arranged for someone in the computer programming department to develop a program to calculate inventory levels and print each revised monthly schedule rather than manually typing them from scratch every month. My plan worked so well that within about two months of introduction, the plant had reduced its production workforce by nearly 20%, and inventory levels in the New Jersey warehouse were already improving.

Metcalf was so pleased by the result that he asked me to visit the Puerto Rico plant with him. The trip went fine until we got to the airport and found that our return seats had been bumped from first-class to coach. There was no way that Metcalf would accept that. He got gruff with me for trying to deal with the ticket agent—"You don't speak Spanish." When she found one first-class seat and I offered to take the coach seat, he snapped at me for settling for less than due. Eventually, our office manager was able to arrange two seats for us, but by that time Metcalf was very cross and even the smallest things got his dander up. Soon after lunch the next day, he called me into his office, told me that I wasn't management material and had to leave. I knew I should grovel, but chose not to. I was given eight weeks pay and a ride home by a technician.

Within the next two months, the other two special assistants lost their jobs, and my former boss Chris Peek resigned and returned to Sylvania. The latter told me that he was very upset at my termination and had decided to get away from a place like that. Dick Wareham from Polaroid called within hours and invited me to return to Polaroid. I preferred to find my own way.

FOSTER-MILLER
November 1975 to December 1978 – Ages 35 to 38

So, now I'm out of a job with eight weeks of pay to tide me over. An early stop was at MIT's alumni placement office which knew all about that strange ECA with its Metcalf and gave me suggestions for handling questions about my termination. I also contacted Foster-Miller Associates (FMA) fairly soon because they had been wooing me for years. But having been chief mechanical engineer for four years and a manager for half a year, I preferred more than a mere engineering job, and that was about what FMA had to offer. After a number of interviews for not-so-right positions over the next six weeks, I accepted their offer of Program Manager.

Foster-Miller turned out to be better than expected and had a lot of MIT people. My boss, Frank and my fellow workers were all interesting. My first assignment Monday morning, November 10 was to manage a US Bureau of Mines (USBM) contract to develop an improved roof support system for longwall coal mining operations. The contract had just been let so the timing was perfect. I needed to learn about longwall mining[10] quickly. One team-member was equally unfamiliar with mining details, but another was an experienced coal miner from Germany.

Soon, the three of us visited a mine at R and P Coal Company in Indiana, Pennsylvania to see a **longwall mining operation** up close. This mine was in a 42-inch coal seam though the main passages were a little higher—like 48 inches high. I had to wear all sorts of unfamiliar equipment: overalls, a hard hat, a lamp on the hat, a cord down the back, a miner's belt, a battery pack on the belt, a self-rescue breathing device on the belt, insulated rubber boots, and knee pads. All this while crouched way over and trying to keep up the pace with experienced miners. The next several days, my legs were so sore I could hardly walk. Descending stairs was excruciating.

I was so overwhelmed with everything being new that I hardly had time to think about what I was doing or seeing. However, while we were crawling across the longwall face, I got stuck at one point because my thigh was too long to fit between the bed of coal on the conveyor and the roof chocks against the roof. When I tried to twist to the left, the battery pack got caught on the chock. When trying to twist to the right, the self-rescuer got caught on the

---

10 In a longwall mining area, two parallel gate entries are mined thousands of feet into the coal seam. Typically, the gate entries are 500 to 2,000 feet apart, each comprising a 3-passage network of room and pillar cuts where each cut is about 20 feet wide. At the far end of this pair of gate entries, a connecting passage is cut to form the face of the longwall coal block. Heavy-duty hydraulic chocks are installed shoulder to shoulder across the face as is a chain conveyor belt and a longwall shearer—the cutting machine. Each cycle of the shearer crossing the face removes about 3 feet of coal which falls onto the conveyor and is carried to the main conveyor in the headgate entry. After the shearer passes each chock, a coal miner operates a sequence of hydraulic cylinders to snake the nearby chocks and conveyor forward about 3 feet to close the gap. Behind the chocks, the roof soon collapses to form a "gob" area.

chock. I'm thinking, "this space is too narrow because *the roof is coming down!*" Meanwhile, the voices of my fellow travelers were growing fainter in the distance and the groans, cracking, and crashing of rocks in the gob area behind the chocks seemed to be growing louder. I was feeling desperate: "Bellows, *What are you doing here?*" I asked myself.

I eventually made it through, and later learned how to walk more comfortably in low coal: Fold your hands behind your butt, hook the lamp cable with a thumb and pull on the cable to lift your hard hat and lamp to face forward. The legs can straighten somewhat, the neck can relax, the chest is stretched for better breathing, and the arms and hands act as a counterweight in back, not a dead weight in front. I even used that technique years later whenever walking in that section of shallow basement in Wayland.

In the end, I visited dozens of mines, several longwalls, many with low roofs, but none as low as that introductory visit. (Please note that all of my descriptions of mining operations reflect a practice seen in the late 1970s. I have no idea of possible changes in the intervening four decades.)

PROJECT – The objective of this **first project** of mine was to design and maybe build a roof support system for each of the head and tail entries that would be more effective than the typical practice of installing supplemental timbers and wood cribs. In addition to the conventional roof bolts, supplemental support is usually needed in the 100-plus feet nearest the face because of the instability of the roof caused by the intentional collapsing behind the nearby longwall face. The original proposal for such a support system was submitted before my arrival at FMA and was developed by people who expected a coal mine to be neat and tidy and thereby accommodate complex articulating machinery without trouble. As we got into the initial contract phase of evaluation and development of realistic designs, including visits to several mines and discussions with even more operators, it became clear that any effective support system needed to be simple, robust, and highly adaptable. (Actually, one of my conclusions after interviewing most of the operators of the 40 or so extant longwalls in the US was that any formalized system such as outlined in our proposal was unrealistic in the complex, irregular, and unpredictable environment of a longwall entry. But this was a government *contract*: it would be inappropriate to shoot ourselves in the foot by admitting the truth and terminating the contract. So we continued to pursue any design that had the slightest chance of feasibility.)

We eventually came up with a moderately good design that appealed to the USBM and to many of the operators I presented it to. Each support consisted of a pair of high-strength H-beams welded together to form a box-section with wings and a hydraulic support at each end that was manually set in place or retracted and folded up close to the beam for movement. As the face advanced, these supports would be sequentially removed by a transporter and reset at the opposite end 100 to 200 feet back. The transporter was a tire-mounted vehicle with a crane arm that could grip the lower flanges of the beams and twist and swing the beam in any direction to negotiate it along the busy and cluttered head or tail entry.

Soon, I also became a **team member of several other projects** or engaged in writing new project proposals, so it was rare for me to work only on one project. After about six months had elapsed, John Curcio fresh out of MIT joined the group, and after about two years our boss Frank resigned and returned to Texas. He was not replaced—we reported less formally to two marketing managers, Bill and Ross, and I took up much of the slack of day-to-day management.

In total, I worked on the longwall contract for nearly three years at which time the beam system had been completely designed and quoted and the transporter had been designed and was being quoted. As my part of this project approached its end,

I was desperately trying to find a longwall mine that would agree to install and test it. I called every longwall mine operator in the country and got answers like: "Looks like a good plan but not applicable to our mine." or "We're just finishing up this longwall and won't be starting another any time soon." or "We don't see a need for it under our roof conditions." One operator finally agreed, but it was rumored that he took it just to get a free transporter. (Well after I left FMA, I heard that the system had been built and delivered, but that the beams lay above ground rusting away.)

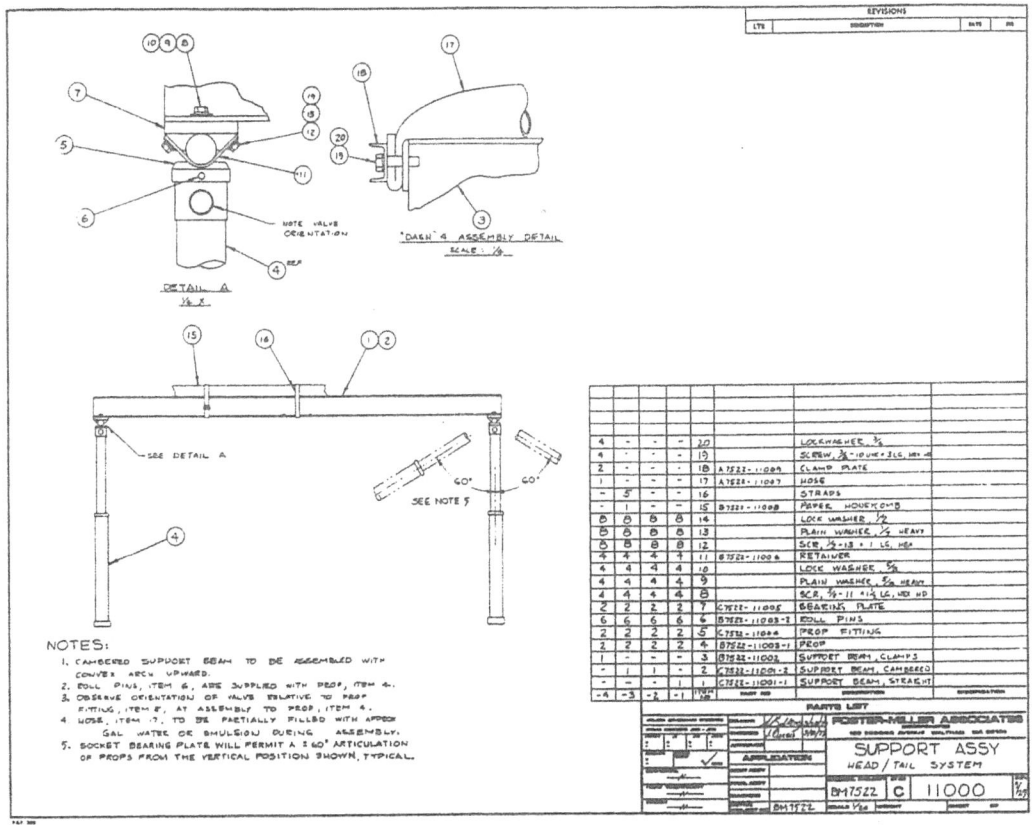

An assembly drawing of the Head and Tail Support System. This figure is a good example of the many assembly drawings I did with drafting machine and pencil for many years.

• • •

Before proceeding to another project, I should mention another opportunity toward my long term quest for financial independence during retirement. During a brief chat with an old friend soon after joining FMA, I learned that Congress had enacted the **Individual Retirement Arrangement** (IRA) just a few months earlier (September 1974). What timing: The IRA applied only to persons working at companies with no retirement plan, and I had just joined one. I took advantage of it and squeezed from my paycheck the maximum of $1,500 per year for the next three years. (For the record, Block had no plan, Polaroid's was generous, and ECA's was useless.)

• • •

One of the other projects I managed from writing the proposal to presenting a finished design was a **temporary face support system** for a roof bolter. In a typical room and pillar mining cycle, first a continuous miner—an enormous machine—uses its rotating cutter to carve an opening 20 feet wide, as high as the coal seam, and 20 to 30 feet ahead. Secondly, the mining machine backs out and moves to an adjacent passage to begin the same process. Thirdly, a roof bolting machine enters this newly cut passage and drills holes 4 to 5 feet up into the roof and installs expansion bolts. The bolts must be spaced every 4 feet across and along the passage. Although there is a heavy-duty machine doing all the work, it requires an operator to stand or crouch adjacent to the drill head and handle the drill bits and insert the bolts. It is the most dangerous operation in a coal mine because the operator is directly under recently exposed and unsupported rock at the newly-cut coal face and is constantly moving the machine from hole to hole. Our project was to design a bolter that moved once for each row of holes and deployed temporary support to the entire area to shield the operator while he bolted the whole row more quickly.

Model of the Temporary Face Support System, a project managed by and mostly designed by me, 1977. This wood model, about 18 inches long, was constructed by John Curcio and me at my home. The cross-slides could be moved to spot the drill head, the supports telescoped, and the four whiffletrees were free to pitch. Note the drive chain in the cross-slide.

During the conceptual design phase of this contract, I visited a model-making shop in Woburn to get a quote for building a model of our bolter. When I arrived in the early afternoon, I saw a lobby full of empty champagne bottles and crumb-covered plates. I asked, "Birthday party?" The receptionist said, "No, we were just awarded a 1.1 million dollar contract to build a 120-foot long model of the depression of the central artery in Boston." That was in 1976—the "Big Dig" was finished 26 years later.

In the end, John and I built that model ourselves in my kitchen and using my table saw. The roof bolter design was a lot more practical than the longwall support. Too bad they didn't continue the contract into construction.

TRAVEL – While at Foster-Miller, I went on a lot of **field trips** to a lot of unusual places. According to an old folder of trip expense reports, I flew on at least 36 trips that included Pittsburgh for USBM and MESA meetings and a number of mines, operators, and manufacturers in the Pittsburgh and West Virginia area; Spokane for meetings with USBM; Detroit, Minneapolis, and Las Vegas for conferences; Lexington, Ky. for mine and manufacturer visits; Salt Lake City for two mines and an interview at Utah U; Denver for mine visits in Colorado and New Mexico; and Charleston, W.Va. many times during our underground testing. On one of my trips to Utah, I visited two mines near Price, and then took a few days off to visit Arches National Park and Dead Horse Point to see the Colorado River and Canyonlands National Park. I stayed in Moab for a night or two and drove to Grand Junction to catch a plane back to Boston.

At one point when I had almost nothing to do, Bill asked me to write a **proposal for underreaming** of deeply drilled holes in the ground. Huh? What does that mean? A consultant was brought in to share with us his knowledge of the practice, and I was told to write it up. I had no idea what to write. I struggled for a couple of days, then went to Bill and confessed my frustration and ignorance of what to write. He calmed me down and said he'd get back to me tomorrow. In the morning he brought me a sheaf of handwritten pages of prose to constitute the basis of the proposal. My only job was to fill in the blanks. Sounded easy enough, but as I began reading his work, I realized that all the hard part still remained. Passages went like: "Foster-Miller Associates proposes _____." "Underreaming can be highly useful to provide _____ in the oil industry." "It might also be useful in the _____ industry." This went on for the thirty or so pages—not a useful fact anywhere. If I hadn't already been unhappy and anxious with the work content at FMA, this turn of events would have been funny. But at the time, it wasn't.

I didn't find contract work very satisfying. The proposals were onerous to prepare and edit into form satisfactory to all the bosses, the monthly reports were often a challenge, time expenditure had to balance the budget, the periodic presentations took substantial time away from real design work, and too often the work content had to be tailored to keeping the contract going despite a weak design. More than once I looked for another job, and got an offer from Polaroid at its Waltham film machinery division. I told Bill of my plans, but he talked me out of it with the enticement of getting their flexible roof drill into working order. Everything about that offer was appealing: managing a full project, freedom from monthly reports, and the challenge of truly accomplishing something. I am certainly glad I didn't go back to Polaroid.

• • •

The **flexible roof drill** shaft and drive chuck (or "drill head") had been designed a few years earlier under contract. FMA was putting its own money into tuning it up in hopes of selling the design or actual machines for a profit. The current status included a lot of pieces, none of which worked quite right, and an irascible but experienced technician, Arthur, who was using a stationary test stand to bore holes in concrete blocks.

THE MAGIC DRILL – The idea behind the flexible drill system had great merit. Roof bolt holes are $1^3/_8$ inches in diameter and must extend at least 4 feet into the rock above a mine. In low coal mines, drilling for 50 inches requires multiple drill extensions and the manual handling of each. There is considerable risk of the operator being injured by either the rotating machinery or falling rock. But a flexible drill shaft could lie horizontally in the machine, curve to a vertical position just below the roof, and become rigid when torqued by the drill head. The drill head was designed to simultaneously grip the shaft, rotate, and push upward by one inch, then release, retract, and repeat. The concept worked and the hardware functioned—at least for short periods.

One of my first challenges was to get the drill head mounted on a machine—otherwise the project would remain a laboratory curiosity. FMA had already purchased a used low-coal face-drilling machine to modify for this purpose. The machine was about the size of a full-sized car except for being very squat. I started by asking Arthur, the project's crusty technician, for his advice. This accomplished two things—one, it got him on my side, and, two, he had a lot of experience and expertise in heavy equipment, so the advice was very useful. I devised a parallelogram lifting arm and bracing foot that he fabricated using a high-strength grade of steel. He installed a large rubber hose as a shaft-way running the length of the machine. Ordinarily, this machine would be called a "roof bolter," but ours did not have the hardware needed to install and tighten the bolts. (See US Pat. No. 4,201,270.) I also ordered 4-foot cubic blocks of "13-bag cement" for our testing instead of using a stack of construction grade concrete blocks.

There was a reported "sealing issue" for preventing the drilling particles from getting into the drill head mechanism. My philosophy was that there was no way of keeping dirt from falling *into* the top of a machine working in a coal mine—better to design a way for the dirt to fall *out* the bottom. That worked.

The hydraulic system had such a serious overheating problem that testing had to proceed for short periods with lots of cooling time in between. I had never studied hydraulic systems, but I did know that "unused" fluid flow was diverted through a relief valve where all the energy that had been supplied by the pressurizing pump was lost as heat. In a pressure system at 5,000 psi, the heat produced was enormous. I completely redesigned the valve system to avoid much of that loss.

The drills consisted of two concentric counter-wound springs with an outside diameter of 1 inch, a length of 8 feet, and welded to fittings at each end. The springs were close-wound of rectangular "wire" $3/8$ inch wide and $1/8$ inch thick, thus there was a half inch hole running down the center. When torqued, the outer spring tried to become smaller, the inner one larger, and the pair locked together to become very rigid. Many of the drills were failing by growing longer with the tight-wound coil turns opening up in spots. There was never a good explanation, but I asked the technician to cut away the weldment at the tail end, and momentarily twist the outer spring to open it up. Presto, the failed shaft snapped into place. No more tail welds—no more problems.

After about six months, we had a **working roof drilling machine** that could be driven around the warehouse and drill holes for long periods without resting or needing cleaning. Arthur rented a steam cleaner, washed down the oily machine, and painted it a typical construction machine yellow. I cut out some gummed black vinyl sheet in the shape of Foster-Miller's logo and applied it to each side and front. By this time, I had arranged for a test site in a 46-inch coal mine at Brady Cline Coal Company in Summersville, West Virginia. Before shipping the machine, I rented a 16-mm movie camera, bought ten rolls of color film and shot 30 minutes worth of action and close-ups of the machine and its articulations in the parking lot behind the warehouse.

In late May 1978 **the machine was shipped to the mine**, Arthur drove his pickup loaded with tools, extra shafts, and other supplies to the mine, and I flew down to Charleston, W.Va. and drove a rented car eastward. The mine was a small one under minimal cover, with the active face only about 2,000 feet into the mountain, and a horizontal entrance directly into the coal seam from a previously worked open pit mine of the same seam. Our test site was about 1,000 feet in and we typically had to walk through water for about 200 feet. The mine was worked by a single shift of ten men, and occasionally I would go up to the face to see real work being done. On one occasion I joined the crew just after they had drilled the face, set dynamite, and were about to blast. The foreman simply wiped the two fuse wires across those conical coil terminals of a 6-volt lantern battery to initiate the blast. The sound of the explosion was largely muffled by all the corners we were behind, but there was a great whoosh of wind and lots of black coal dust stirred up.

The first week was mostly taken up with some last-minute preparations, getting our site established, connecting the huge power cable to an explosion-proof electrical panel, and getting the machine to the site. Typically, Arthur posted himself near the drill head, spotted the drilling site, and directed operations while another technician sat in the operator's seat, manned the controls, and recorded each drilled hole in a notebook I had prepared for the purpose. It was my intention to be there for the first week to make sure testing was going smoothly, but at the end of the first week, it was clear that I needed to stay for the second. There were enough problems to resolve each week that I ultimately stayed on site for the entire six-week period it took to drill the goal of a thousand holes. Somehow, I enjoyed every minute of it! There is something about mining that is very alluring, and I also recall that in all my visits to coal mines, I never encountered a single miner who didn't love his work.

During one of those weeks, I rented the movie camera again along with lights and shot ten rolls of underground operations. This was much more difficult than the first ten rolls because of the tight and messy work space, but eventually some good shots were obtained.

Generally, we entered our work site after the miners had already begun their day. I usually came to the surface for lunch, and my favorite exit route was to go deeper into the mine, avoiding the water, to a place where I could crawl under a brattice cloth into the passage with the rope belt for removing the cut coal. The foreman had taught me how to jump onto the moving coal-laden belt, lie on my stomach, grab the edges for stability, lift the edge slightly to avoid pinching my fingers at every roller, then jump off at the critical moment before being dumped onto the coal pile. Most miners ate below, so lunch was usually just us, the foreman, and one or two others. After lunch, it was another 1,000 foot walk crouched over and through the section of water. At quitting time, I exited the short but wet route.

We were at that mine from May 17 until June 22, 1978, drilled 1,215 holes with a total depth of 5,217 feet, didn't damage a single flexible drill shaft, and one of them was used for over 725 holes before we switched over to trying a different shaft.

Later in the summer, we took the equipment to a T&T Coals Company mine near Morgantown where the roof consisted of very hard sandstone. That was a robust test that we ran for only two weeks. It was during this cycle that we used and tested some silicon-carbide bits supplied by GTE Laboratories. (By coincidence, GTE approached FMA to design and build a rock-drill testing machine. As a quick test for GTE, I proposed that GTE supply us with drill bits and we would test them in real rock using our flexible roof drill. Several months later, this collaborative connection probably helped with my landing a job at GTE.)

When I first began **making movies**, it was with the simple plan of splicing together a short series of shots for historical record. My first editing cycle simply selected the best of any multiple shots and put those scenes in a logical sequence. The result ran for about 20 minutes and was pretty boring. However, the marketing VP liked it enough that he asked me to edit it down to about 5 minutes and add a narrative sound track. Sounded like fun and should be easy. I could not have imagined how much work that movie would turn out to be.

First challenge was to decide which of those perfectly good shots would be sacrificed. Even that agonizing cycle fell short of the 5-minute goal but eventually it was done. When reasonably satisfied, I wrote and polished a narrative over many cycles, encouraged the in-house artist to make a few animation gels to explain the flexible shaft and drill head, composed titles, and hired a local radio announcer to read the narrative.

At the film studio, I learned all about the A and B movie roll production system and the noiseless splicing of the sound roll. Eventually, we had our 5-minute movie. I thought the project was done, but then was asked to have the 16-mm movie made into an 8-mm movie mounted in a continuously running cartridge so it could be set up to run all day at our booth at the Las Vegas mining show. A great experience, but I wouldn't want to be in the movie-making business—I would never have believed how much work was required to make a five-minute movie. It is no wonder there are so many credits at the end of a typical Hollywood movie.

During that fall after the summer tests were done, I entered into several cycles of on-again, off-again interviews at GTE Labs. Late in the fall, GTE got serious, ran me through about ten interviews, and ultimately offered me a job. The roof drill was still not an off-the-shelf product, but I had surely brought it a long way and had prepared a complete set of detail and assembly drawings that could be used to produce more copies of the machine that we successfully tested. In the year or so that I kept in touch, it seemed that FMA was not aggressively moving that project ahead. I left there in good stead, didn't burn any bridges, and even did a little consulting for them some years later.

ADULT EXPERIENCES, TALES, and ANECDOTES

This chapter, like its companion, *Youthful Experiences, Tales, and Anecdotes* found earlier, is a collection of various topics and stories with no other place to fit. Some, like the Zundapp problems, furniture building, and plumbing refinements connect to my life as an engineer. Others like the photography workshop and silversmithing are only peripherally technical. And a few, like the summer trip, the mouse story, and service to my community are extraneous adventures that have been especially memorable to me.

My Family – I first married a woman I met through Polaroid. I was 29 and it was my first year after moving out of that bachelor pad on Hayes Street. We had a daughter three years later. I was totally unprepared for my reaction of feeling instantaneous love and protection for this tiny person. I remember holding her and watching her move and thinking she's like having a doll that *works by itself*. The experience of bringing her up and watching her grow each day was a real joy. When she was about three, she took up the practice of running and jumping onto my back whenever she saw me squatting for some kind of work on the floor or ground. It usually caused me to topple backwards and interrupted my pace, but I surely missed it when she stopped.

In our seventh year, the marriage fell apart, and I was devastated at the prospect of losing my daughter. Fortunately the joint custody separation was not bitter, and her new home was conveniently nearby for frequent pick-ups and drop-offs. It was doubly hard when her move away occurred a few days after I lost my job at ECA.

Three years later, I married a woman introduced by my boss at Foster-Miller a year earlier. She had two boys respectively two and three years older than my daughter. During our courtship, she was working at Amica full time, rearing her sons alone, self-marketing her house, and seeing me frequently. She had a keen sense of humor, finished my sentences, did beautiful needlepoint and other crafts, challenged me on many issues, and was a good organizer. We had a lot of shared interests, met each other's friends, worked on our houses together, and spent time with our kids. She gained a daughter, and I gained two sons.

Three years after our marriage, we had a son. Again, I developed immediate love and protection for this little boy. He was a happy baby and child with a quick smile, and every day with him was a delight. When he was about three, he learned to get himself up each morning, come around to my side of the bed, wake me up with a big smile, and then go downstairs to watch TV. I didn't always like being awakened, but I was truly sorry when he no longer did so.

The children are all grown now, and three have families of their own with two children each. We have five grandsons and one granddaughter.

My father died at 85, my Mother lived to 100.

The Zundapp – In May of my senior year at MIT, I bought a used Zundapp motor scooter from one of my fraternity brothers. Some of my escapades with it appear elsewhere. Failures were *always* happening with that machine, and I repeatedly had to repair or replace some part. I soon learned where to get repair parts and routinely carried wrenches, pliers, screwdrivers, tire irons, tube patch kit, hand-cleaner, and rags in a toolbox that I had installed behind the seat.

The troubles were so frequent that at some point I began keeping a list. That list with over thirty entries recently surfaced and included: leaky carburetor; wires breaking; many flat tires patched; dead battery; adjustments or repairs needed for horn, brake light switch, chain, gear shift, tail pipe, *etc.*; and replaced parts included cables, chain, brake light switch, ignition switch, battery, and rear brake drum. Early on it had a serious starting problem that I eventually traced to an intermittent and elusive short at the magneto. The engine seized three times leading to my rebuilding the engine out on a Boston street and replacing the piston, rings, and gaskets. And of course there were many repairs following my one accident—I was rear-ended.

The motor scooter was a wonderful machine to have had in Boston and Cambridge. I could usually get around traffic, park almost anywhere, and it zipped me along as fast as or faster than most of the cars around me. Although it was riddled with troubles, I got a certain pride from being able to diagnose and repair most of them with dispatch. I missed it after selling it.

My Sketching and Artwork – I was never very good at drawing pictures of people, and in the rare instances that I drew them, they resembled stick figures. Most of my early drawings emphasized their mechanical aspects. If drawing a party scene, I would get those shoes just right, the light fixtures and the cake drawn precisely, and the chairs drawn in detail. But my drawings of people were just not realistic. In time, I began to use a straight edge for architectural drawings or to lay out some project I was planning.

My early line-work was dedicated mostly to architectural drawing. I looked forward to the Sunday newspaper, and while Bill read the sports section, I would turn to the real estate section and look for the latest house plans to study both the design and drafting technique. I noticed that architects drew their corners overlapping, but I could never bring myself to be so *sloppy* no matter the convention. My corners were always closed.

For some class in high school, I volunteered to draw a poster-sized bank check and explain the proper way of filling out a check. I drew this giant check using India ink, various sizes of Speedball™ pen nibs, and a ruling pen for the fill-in lines. Also in high school, *Scientific American* had an article on origami, an art-form little known in America at the time. I learned how to make a flapping-wing bird from a square of paper and later enjoyed making a number of other origami pieces described in various sources.

Apparently, I had the ability to draw perspective views of mechanisms before college but developed it to a greater extent when at Polaroid. I was impressed by my boss's ability to sketch out a perspective of an idea of his and noticed how much more effective it was than an orthographic projection for conveying an idea.

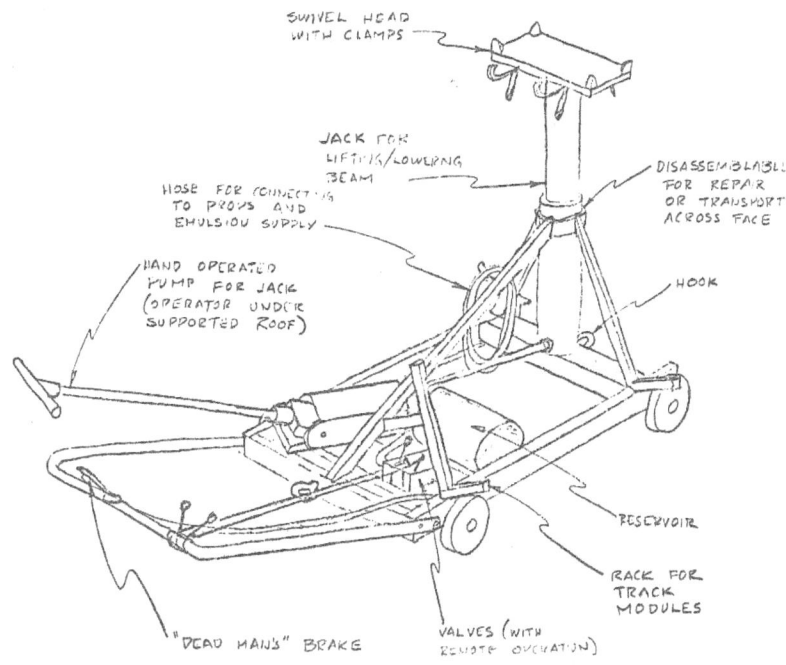

One example of many hand sketches I prepared for designers, drafters, project team members, or reports. This one was drawn at Foster-Miller to explain a proposed concept.

This newly-refined skill was a godsend at ECA. Although I had a drafting board in my office and used it often, I also had so many overlapping projects that I was always handing-off something to the design group. I usually had one, sometimes two designers working exclusively for me with other drafters assigned whenever needed. This is where I really found that a well-executed perspective sketch of an experimental device, a production tool, an assembly fixture, a testing instrument, or part of a new product was invaluable. Two or three hours spent on such a sketch might keep the designer busy for two or three weeks with little or no follow-up. It was amazing. The assembly tooling, fixtures, and test instruments for the LED scanner took only a day or two to think through and sketch out, but the finished hardware ultimately served an assembly line of about 35 people in the Puerto Rico plant. And as mentioned earlier, I sketched the perspective cutaway view of Fig. 1 in US Patent 3,825,913 for ECA's extended scanner, also seen at the lower right of the front cover.

At Foster-Miller where much of my work product was proposals and reports, I often made such drawings to incorporate as figures. At GTE, I used sketches for two purposes: guidance for designers and drafters and figures for those many reports.

I rarely applied these skills to art *per se.* I consider much of my photographic work to be art, but that didn't require hand-eye coordination. Several times, I took the time to hand-draw carefully-implemented maps to people's houses that included a slightly tilted magnifying glass for the enlarged inset and India ink applied with various pen widths. I undertook calligraphy while in college and continued to use it for various purposes since then. I calligraphed invitations or the artwork for several weddings

A typical 5x7 photographic Christmas card contact-printed by me, this one in 1967. Photographed by me using the Calumet 4x5 view camera and Plus-X film.

For over a decade starting in grad school, I made photographic Christmas cards. After I had my own darkroom and especially when using 4x5 film, I refined the Christmas card-art to use a favorite artistic photograph along with a greeting that I calligraphed then converted to a Kodalith negative for contact printing. At one point, I was making and sending about 50 each year.

When computer-aided design (CAD) software came along, I was an early-adopter, but I always missed and still do miss taking a pencil to paper and creating a precise and carefully-executed drawing by hand. I miss the pencil in my hand; miss the instruments for making and changing lines and circles and other shapes; and miss figuring things through calculations or descriptive geometry projections rather than asking the computer for the answer. And I miss the satisfaction of keeping my drawings cleaner than most other engineers, designers, and drafters could.

Also, I am certain that inventions, of which I have a bunch, are a product of creativity, and therefore a form of art.

Drawings and Blueprints – When I first went into an architectural office I saw blueprints—copies of drawings that were literally blue. The entire background was a dark blue and the lines and printed notes were white. Later I learned that they were made with a wet process and had to dry before folding or spreading out on one's desk. The archival copy of my two theses required blueprints, but other than those, I never had a true blueprint made from one of my drawings.

I first saw a drawing copy that was not a "negative" blueprint in my teens. The lines were dark blue and the background was near-white but textured with a haze of blue. It had been made with the ozalid process, a dry process that used ammonia fumes to develop the image and was much faster than making a blueprint. Both of the above processes required the drawing to be translucent and the copy paper to be placed directly beneath the original drawing while running through the print machine.

In high school, we used a heavy salmon-colored paper, which could not be copied except by Photostat.[11] The drawings I did in fifth grade at the architect's office were on drafting vellum, a robust but translucent paper. For all my drawing classes in college, I used vellum and made ozalid copies—often referred to as *prints* or sometimes *blueprints*. Every engineering office I worked in until about 1990 had a print machine, drafting boards, large sheets of vellum paper, rolls of vellum, and large, shallow-drawer plan-files. Most companies used preprinted sheets with a border all around and a custom title block in the lower right corner. The larger engineering offices had a technician assigned to do nothing but keep originals filed in an orderly fashion and make prints.

Computer-aided design systems came into being concurrently with large xerographic copiers and printers, so the previously-common ozalid machines quickly became obsolete in the 1990s. After that, I never had easy access to printers larger than 11 by 17, so I quickly adopted the custom of arranging my CAD drawings to fit on one of the two smaller sizes.

Speaking of early technologies, 30% of my career was performed without a calculator—nothing but pencil and paper calculations using my head, a slide rule, or log tables. Over 55% was without ready access to a computer—all drawings were done by hand and all reports written in longhand before being typed by a secretary. And 75% of my career was without a cell phone.

Furniture Making – My instinct to design and build beautiful things seemingly goes back a long way. The first piece of furniture that I clearly recall making was a pine wood chest made when about ten, and still in my possession. It measures 25.9 by 11.5 by 11.5. I made it of ordinary pine (maybe yellow pine) and was very particular about the detailing of the end panels. The panels were boards, but they were framed with two-inch stiles that I chamfered with radii that tapered out just before the junction with the adjacent stile. The stiles served as lifting handles, and I stained the chest mahogany.

---

11 Before the Xerox 914 was introduced in 1959, copies were rare and expensive. *Photostats*™ were made by photographing a page directly onto thin photographic print paper, developing, fixing, and drying it with the result that the text was typically white and the background black or nearly so.

In addition to the tall three-legged stool, one-legged stool, redwood lamp base, photographic print box with inlaid glass and articulated cover, enlarger, and electrical test box already mentioned elsewhere, some of the more interesting or unusual pieces I designed and constructed over the years include the following items.

My grandfather and I worked together to make a utility table for Mother. It has spindled wood legs that we got from a discarded table, plywood for the top and two shelves, and casters so it could be easily moved. The middle shelf is attached using a typical method of Old Man's: a block nailed to the plywood, and a stove bolt running sideways through the block and leg. It was rather crude but quite serviceable. I now use it in my shop.

When in the ninth grade, one of my courses was shop. The instructor gave each of us the option of building something of our own choosing. Mine was to be an oak bookcase. I learned to cut dados and rabbets and to scrape the wood to remove the planing scallops. It was assembled with round head brass screws. I finished it at home over the summer.

For photographic storage, I made a pair of covered boxes to serve as trays for the 2 by 2 color slides. I still have these, which measure 2.5 by 8.1 by 2.6 high. Each was made as an enclosed box then sawn apart with a handsaw to create a base and cover. The joints are not beveled but are glued with my then-favorite Duco Cement and reinforced with brads. There are three tiny metal pins in the cover that mate with holes in the base to align the two.

The print-, test-, and slide-boxes were made mostly from a stash of $3/16$ thick walnut-veneered plywood also used for many other projects. I made an effort to select good sections of grain and orient it aesthetically and consistently. Those items were never varnished.

Also, there was the bookshelf I built as a freshman at Sigma Chi, and once ensconced in my first bachelor pad, I made that coffee table and low bookshelf already described. Those were varnished with polyurethane, a then-new discovery of mine, and one that also protected many other furniture items from drink rings and water spills. For my photographs, I made about a dozen picture frames entirely from scratch shaped and cut from a lauan board.

Soon after moving to the Cambridge apartment house, I built a four-legged platform for a double mattress for my then much-larger bedroom. Also, I built a dining table of heavy rough yellow pine that I purchased at Dwain's Wrecking warehouse as salvaged beams. The wood was so hard that drilling holes was like drilling though pine knots for both difficulty and that wonderful pine aroma. Once I had purchased and set up a table saw, I built a five-drawer chest of drawers of oak and a coffee table from solid walnut.

When my daughter was three, I designed and built her a dollhouse. It was finished before the two of us went to Charlotte for Christmas and upon her return, it was where "Santa had left it" by the fireplace. I remember her spread hands flying to each side of her head and exclaiming breathlessly, "A dollhouse!" Later, I made a wall-mountable dollhouse architecturally similar to the Belmont house. She decorated and played with both for years.

In Belmont, I built a coffee table using rough studs that I had removed from the kitchen renovation project. That table was used in the family room in Wayland for nearly thirty years and ended up in my son's Cambridge condo. He and his wife now have the 30 by 48 oak butcher-block table made with well-inset pedestal legs that we used for 28 years for kitchen dining. I also assembled two kits my in-laws gave me: a walnut butler's table and a folding walnut side table with oval top.

Equation of time – A friend and I once got into a discussion regarding the fact that although the shortest day of the year is typically December 20, the earliest sunset and latest sunrise are different by about ten days. But neither of us could remember which way the two shifts went. I looked it up, found the answer, and devised an easy mnemonic:

> The earliest early is earlier.

Explanation: The *earliest* sunset, which is the *early* event in winter, is *earlier* than the winter solstice, therefore about December 10. Similarly, the *earliest* sunrise, which is the *early* event in summer, is *earlier* than the summer solstice, therefore about June 10. The opposite is also true: *The latest late is later*—the 30th of those respective months.

The Big Summer Trip – Several factors converged to convince me to spend the summer of 1964 exploring the United States. This trip was not a technical or engineering part of my life, but it was such an important slice of my life at the time, and it gave color to so much of my life for years to come that I must include it for perspective. A short summary of that trip follows.

During the spring of my last year at MIT, I bought a three-year-old VW beetle; bought camping supplies; gathered my two cameras, three lenses, and 28 rolls of film; joined AAA where I picked up guides and maps; and packed a minimal supply of clothes and $700 in travelers' checks.

I left Cambridge with my girlfriend on Thursday, June 4, 1964 and followed this outline to visit these highlighted places. Through Connecticut and New York to the '63 Worlds Fair on Long Island where she and I said goodbye for the summer. Down the eastern coast through New Jersey, Pennsylvania, the Delmarva Peninsula, and North Carolina to New Bern, Camp Sea Gull, and Charlotte to visit my parents. Then southwest through South Carolina, Georgia, Alabama, and Florida to Ft. Walton Beach.

From there straight westward through Alabama, Mississippi, Louisiana, Texas, and New Mexico to see Carlsbad Caverns National Park. Then I wandered through New Mexico, Arizona, Utah, Nevada, and California to explore White Sands, Petrified Forest, the Painted Desert, Walnut Canyon, Oak Creek Canyon, Sunset Crater, Grand Canyon, Brice, Zion, Las Vegas, Hoover Dam, Joshua Tree National Monument, Mount Palomar, and reached the Pacific Ocean at Oceanside. Then northward through central and coastal California, Oregon, and Washington to see Sequoia, Yosemite, San Francisco, Muir Woods,

Redwoods, Crater Lake, Olympic National Park, a climb up Mount Rainier, Seattle, and into Canada east of Vancouver.

Then it was northeast to Lake Louse, Banff, Jasper, and south to Waterton Lakes and Glacier National Parks. Then south through Montana and Wyoming for Yellowstone, Teton, Yellowstone again, and diagonally northeast across Montana to Williston, North Dakota to visit new friends. Then south through North and South Dakota to the Black Hills, Mount Rushmore, Badlands, and east to Mitchell to have the car repaired, through Minnesota, Iowa, and Wisconsin to Milwaukee. From there, I headed north through upper Michigan, into Canada at Sault Ste. Marie, around to Toronto, back into the US at Niagara Falls and across New York and Massachusetts to home on August 24.

It was a delightful and unforgettable trip that includes memories that come back to me on a constant basis. I explored 15 National Parks plus two more in Canada, saw innumerable classic sights, logged 13,124 miles, hiked for more than 122 miles, spent 75 days on the road, slept under the stars 65 nights in a row, and spent $510.75 including $146 for gas, $52 for car maintenance, 99 cents a day for food, and the rest for fees, clothing, and other supplies.

Yosemite Workshop – In 1966, my second year at Polaroid, Meroe Morse recommended that I go to the Ansel Adams Yosemite Photography Workshop. I had bought a 4x5 Calumet view camera the previous year, was well-versed in its operation, and the timing for some artistic training seemed perfect. Besides, it was on Polaroid's nickel, although my boss, feeling it was not in line with my engineering design work, made me use vacation time for those two weeks away. But what a vacation: Yosemite Park, warm weather, Ansel Adams, taking pictures, interesting associates.

There were nearly 60 photographers attending. Ansel had five or six assistants who instructed and helped out in the field. We often started a day in a meeting room at the Best Studio with a talk or discussion before venturing out. Some talks were held outdoors on site and informally. Ansel talked in detail about the *zone system* of exposure that he largely invented, or at least popularized years before.

Ansel and his wife Virginia had a home behind the Best Studio. *Best* was Virginia's maiden name, and her family had operated concessions in Yosemite for many decades, perhaps owning the property. My association with Polaroid helped to single me out, and several evenings Ansel invited me to join him, Virginia, the assistants, and occasional visitors for dinner. Dinner was usually outside on the patio behind the house and near the darkroom, a standalone building. He was a very expansive character and encouraged lots of talk and laughter during these informal gatherings. Though he had lots of stories to tell, he did not dominate the parties. Virginia was more reserved and quiet but very caring and conversational at times.

On two occasions, Ansel played the piano for us. As a youth he had studied to become a concert pianist before he made the difficult decision to devote his time to photography. Despite his arthritis-gnarled fingers, he played beautifully with a lilting and gentle touch.

Ansel Adams in 1966. (Photo by the author using a 4x5 view camera while attending Ansel's Workshop.)

THE EVER-POPULAR: MOONRISE – During one of his talks, he told the story of *Moonrise, Hernandez, New Mexico*. At the end of a frustrating day, he was heading home when he came around a curve and saw the scene captured in *Moonrise*. This was a must. He stopped the car, set up the tripod and 8x10 camera as fast as he could, and, while returning to the car for a film holder, he calculated a tentative exposure based on placing the moon's known brightness of 320 candles per square foot onto Zone 7. He set the shutter, inserted the film holder, pulled the slide, took a picture, and as he was pushing the slide back in, the clouds behind him closed up and darkened the cemetery. There was no chance to use an exposure meter for a second shot. But he used the meter to determine how close he had been and found that the foreground was "vastly underexposed." [12]

But Ansel, the supreme photographic technologist, had a method for correcting that mistake—the water-bath technique. Using this technique, he first put the negative into developer, agitated it for about 10 seconds, then moved the negative to a tray of water and let it sit for many minutes without any agitation whatever. Then he moved it back to the developer and repeated the process.

He said that he kept up this cycle for about six hours (probably an exaggeration). During the 10 seconds of agitation in the developer, the developer would fully soak

---

12 His method of calculating exposure: If you know the brightness in candles per square foot of the average scene, or of Zone 6 if you are Mr. Adams, and you know the ASA number of the film being exposed, the calculation is that the shutter speed is the inverse of the ASA number and the f-stop is the square root of the brightness. You can walk your way up or down the f/-stop vs. shutter speed scale as needed. For the *Moonrise* case, he wanted 320 candles per square foot to be on Zone 7, so Zone 6 would be 160 candles per square foot, the square root would be 12.6, leading to an aperture of about f/13. Assuming that his film was ASA 25, the shutter speed would be $1/25$ second. Knowing his preference for small apertures but not needing to overdo it in this case of mostly distant scenery, he might have opted for $1/25$ second at f/11 or the equivalent of $1/10$ second at f/16.

into the emulsion. During the many minutes in the quiescent water, the developer in the fully-exposed regions such as the moon and white clouds would quickly exhaust itself because of the large amount of silver halide that was ready for development. However, the developer that was absorbed into the emulsion in the underexposed foreground had so little silver halide needing development that it was not immediately exhausted and continued to do its job of developing as long as the water was not agitated to wash it away. Through this technique, he was able to develop the highlights (moon, clouds, and gravestones) for a total of several minutes, while the shadows (the foreground and fields beyond) were developed for several hours.

Ansel added that even with this compression of the film's dynamic range, the negative needed a lot of work during printing. Typically, he would dodge the foreground as he printed so that it did not become nearly black, and he would burn in some extraneous high clouds to emphasize the darkness of the sky. As a result, there are many different versions of *Moonrise* from one print to another.

Even after the Workshop, Ansel Adams and I kept in touch through his consultancy with Polaroid. He seemed to think I might be a more open ear than whoever was his former contact person, and directed many memoranda to my attention. We called one another periodically to discuss these, and I distributed them as seemed appropriate. I don't believe that many, if any, ever resulted in product changes at Polaroid. When I was president of the Polaroid Employees Photography Club, I invited Adams to speak at one of the meetings. It was well advertised, and we had a big crowd show up.

After meeting Ansel Adams and with the perspective of knowing Edwin Land, I decided that they were both "great men." But, what made them great? I concluded that greatness was defined by three characteristics: competence; creativity; and exciting and giving human qualities such as warmth, patience, interest, and enthusiasm. Both men displayed these three aspects in their character. They exemplified greatness in many ways: their profession, their outside interests, their friendships, their service to community, *etc.* I was blessed to have known such greatness early in my life.

The Yosemite Workshop was a fabulous experience in many respects, but sadly, it had one deleterious effect on me. I stopped making photographs. True: I didn't stop altogether, but my interest waned quickly, and I found it hard to go out for a day of photographing or to come home for an evening of developing or printing—activities that had been such a rich part of my life for the previous several years. I believed at the time that it was caused in part by the recognition that my own talents were too far down the scale from the veritable champions and that I could never hope to accomplish as much as the masters. That alone should not have stopped me, but somehow I felt unwilling to put so much more work into something that had been a quite joyful avocation, but one with which I had come to feel so much inadequacy.

A year or so after my attendance at the Adams Workshop, Mrs. Land, in appreciation for my house-sitting their home on Brattle Street, offered to have Ansel print a photograph for me. Because of the duality of its beauty and the technology behind it, I chose *Moonrise*. I had no idea it would be so large: the print was 40 by 30 and was mounted in a simple but elegant white frame. Ansel knew he would be printing it for me when Mrs. Land ordered it. Years later when Ansel was visiting Polaroid in 1979, I took the mural into one of Polaroid's labs, saw both him and Dr. Land for the last time, and had Ansel sign it for me. He signed the back side "for Alfred Bellows" with a felt-tip pen.

Second Trip to the Grand Canyon – In September 1967 while working at Polaroid, I got a call one night from my parents who were traveling in the southwest. Dad suggested that I join him to climb in and out of the Grand Canyon. I rang off the phone, called the airlines, arranged a flight for the next afternoon, and took a couple of days off from work. Meanwhile, Dad arranged for a mule driver to meet us at Indian Gardens, a point about half way out of the canyon so that he wouldn't have to endure the entire, grueling trip back to the rim. What a delightful spree. First day we hiked 7.1 miles down the same Kaibab trail I had descended three years earlier for a night at the Phantom Ranch with a served dinner and clean sheets. Early next morning, we climbed five miles up the Bright Angel Trail in the relatively cool air, took a rest and water stop at Indian Gardens, and met our private mule-train ride to the top. Mother, Dad, and I dined together, and the next day I returned to Cambridge.

Silversmithing – In the 1967-68 season, I took a beginner's silversmith course. It covered the basics of shaping a sheet of sterling silver with a hammer, how to periodically anneal it for additional shaping, and how to silver-solder parts together. The instructor also briefly explained the making of wire shapes and advanced shaping techniques, but these were not covered in the course. The shop included a large collection of hammers and stakes to work with.

My friend Len, the versatile model maker at Polaroid, took an interest and suggested I should have a *hallmark* tool. I designed a stylized interlaced AHB surrounded with a rounded trapezoid. Using a pantograph engraver in two stages, he greatly reduced its size to make a tool with recessed letters that would form raised letters in the silver when the tool was struck into the silver object being marked.

My second sterling silver project patterned after a Japanese ceramic bowl. This bowl is six inches in diameter, 1.5 inches high, and weighs 5.4 troy ounces.

I ended up making three major pieces and several smaller ones, including some earrings. The first was a small 1.6 by 2.5 inch oval salt seller with a base soldered in place. The second started with a 6-inch diameter flat of silver that was worked into a shallow cone shape with hours of pounding against a wood stake with a cylindrically-shaped hammer head to push that point down. My goal was to recreate a beautiful Japanese ceramic bowl I had seen at the Honolulu Academy of Arts in late 1967. At the time, I had been so impressed with its shape and delicacy that I had sketched it as well as possible through the glass case. The important features were the angle of the cone-shaped bowl, the smallness of the base relative to the bowl, and the angle of the conical base. My reproduction in silver was quite successful. As I approached completion, the instructor showed me how to broaden the edge slightly, similar to the raised edge of a typical coin, to give the silver piece a nice finish.

The third project was a baby's cup. This had to be made in three pieces: the side rolled from a rectangle, soldered with the highest melting-temperature silver solder, and hammered to shape; the bottom soldered in place using a lower melting-point solder; and finally, the handle. It was beautiful.

Another thing the instructor taught us was to finish the bottom—in other words, shape and polish the hidden parts as well as the rest of the piece. Although that might be a small part of silversmithing, I adopted that instruction in a broader sense to be a metaphor for the optimal approach to many engineering and social situations. Applying finishing touches to the seemingly unseen parts of a project has thereafter been one of my mantras.

Scopons – While at Polaroid, I coined the term "scopon" to refer to the path of a light ray, but in the reverse direction. When describing complex optical systems, it is often easier to describe the path of light as it bounces off of mirrors or refracts through lenses and prisms as though it traveled from film to model or from retina to girlfriend. In fact, Dr. Baker, our consultant and the designer of the first U-2 spy plane cameras, often used the expression "tracing the light rays backwards" when describing one of his designs.

Starting with the Greek *skopion*, "to view or examine" and putting it in the same form as *photon*, I chose to call this imaginary effect a *scopon*. In fact, I believe people *think* this way too: "the eye looks out to see the distant horizon." Have you ever heard anyone say "the light waves from the distant horizon radiate in all directions and those that happen to fall on the lens of an eye are focused onto its retina"? Baker liked my term and thereafter used it often in describing his latest design. Land also took up using it, but it was usually accompanied by a twinkling glance in my direction.

No one else seemed to take me seriously, but I still believe scopons to be light-like particles that arise from the retina, film, photocells, CCDs, *etc.*, and project outward through optical systems.

The Only Mouse I Ever Admired – One night I went to my workshop, which was approached through a dim passageway, and the shop itself was opaque with darkness. I reached in and flipped on the lights to create fluorescent brilliance. On this occasion, I got a surprise—a mouse was frantically scurrying about on the floor a few feet in front of me. In an instant, I was so frightened by the frenetic movement before me that I wanted to get up on a chair.

Within a second, I gathered my wits enough to know that I certainly didn't want to get up in any *chair*. In fact, I was so incensed by being made to feel like the classic frightened woman that I became very angry at that mouse. I stomped my feet and shouted obscenities at the top of my lungs. I continued to stomp, slowly approaching him. That little mouse continued to scurry about just in front of me, and I continued to stomp my feet and shout. This was all taking place within a few seconds after I had turned on the light.

Very suddenly, the mouse died. He had a heart attack and just died right there in the middle of the shop floor. He tumbled over on his back with his little feet up in the air. I could see those little paws with those tiny fingers collapsing closed and drooping in the complete relaxation of death. For a moment I felt sad. His little body was lying there with feet up; he looked so vulnerable. He *was* vulnerable—I had just frightened him to death.

But after my brief reverie, I remembered the extreme fright he had caused me and realized how indignant he made me feel. It then occurred to me that maybe he was *playing 'possum* in hopes that I would abandon the pursuit. So, just to make sure my vengeance was complete, I looked around for a weapon. Weapons were everywhere in my shop: hammers, wrenches, and scrap wood for clubbing; screwdrivers and files for stabbing; nails and screws for shrapnel; or boxes and buckets for capturing. But my weapon of choice was a brick. Very slowly, I stealthily reached for a brick on my workbench. I picked it up and slowly adjusted it to be directly over the dead mouse. I carefully surveyed the line of dropping to be sure. When I was absolutely certain the brick was directly above his little body and perfectly square to maximize the impact, I dropped the brick.

Well, the brick was about half way to the floor when that dead mouse suddenly came alive and jumped to his feet. Those tiny fingers were not truly relaxed in death, rather that mind was intently watchful and finely-tuned to survival. That mouse had also figured out that he better run in a straight line this time, not in circles. Before that brick had even reached the floor, he was all the way across the room finding a hiding place in my stack of wood scraps.

By that time I was so amused and charmed by his cleverness that I felt he surely deserved to live. I hope he lived a long and good life.

Plumbing Orthogonality – My early mentor, Dr. Land often challenged us to think *orthogonally*. Orthogonal means at right angles, and as he used the expression, he meant us to think in unconventional directions—a paradigm that is also captured in the modern expression "think outside the box." (In 1962, Land delivered a commencement speech at Stanford entitled *Orthogonality*.)

In the 1980s, I chose to apply such an approach to some of the hot water pipes in our house with excellent results. Years later I applied it even more aggressively at two other homes.

> **UNCONVENTIONAL PIPES** – One of our second floor bathrooms had the dual defects of having to wait over a minute to get hot water to the sink or shower and the risk of getting scalded if someone flushed a toilet or ran the washer. During the kitchen renovation on the first floor, I planned an unconventional, "orthogonal," approach to obviate both defects. A much shorter routing of the pipes would improve the former, but my hypothesis for eliminating the latter was to dedicate a hot-cold pair of pipes to the shower and to source those pipes as close to the outlet and inlet of the water heater as possible. Thus, any pressure changes in the water supply should equally affect both of those pipes. When executed, the plumber also installed undersized $3/8$ inch pipes to hasten the arrival of hot water still more. (Actual

dimensions: 0.38 ID, 0.50 OD, and a third such pipe for the sink's hot water was included.) The result was hot water arriving in about five seconds and no adverse affect from using water elsewhere.

I specified similar changes for three other showers in two houses. (Please note: These diameters of less than half-inch pipe do not meet the plumbing code. I had satisfied myself by previous experiments that flow rate would be equivalent, demonstrated that to our plumber, and sweet-talked him into meeting my specs. The pipes were then covered with half-inch foam pipe insulation to both hide them and minimize the heat loss.)

For improvements at another house two decades later, I specified even smaller lines than for the earlier baths: $3/8$ tubing (*tube*, not *pipe*, is 0.31 ID, 0.38 OD). Moreover, he installed the tubing to follow the most direct path possible, meaning at an angle across the basement. I determined that the speed of getting hot water to the kitchen sink dropped from 35 to 8 seconds while the full-open flow rate decreased by only 7%. Similarly for the master sinks, the speed dropped from 42 to 10 seconds while the flow rate dropped by 5%. For the shower, the speed dropped from 37 to 13 seconds with no noticeable decrease in the flow rate.

In summary, undersized pipes installed directly to a single sink or shower can deliver hot water much faster, with little reduction in flow rate, no affect from flushing toilets, and a quick transition from cold to fully-hot because the small diameter doesn't allow much mixing.

Wayland Swimming and Tennis Club – Even when volunteering I gladly take on projects to design, repair, or build. We joined WSTC in the spring of 1987. That membership was a boon for our son, who was nearly six. He quickly joined the swim team, which began a sport he continued in one venue or another through college. He also spent long days at the club playing various games with his friends.

Within the first year, I was talked into becoming Buildings and Grounds Manager. In that position, one of my jobs was to plan the four Saturday morning work sessions in April and May when each member could put some sweat-equity into the Club. Most of that was routine: raking leaves, resurfacing the clay courts, cleaning chaises, spreading mulch, *etc.* But there also were special projects: mending fences, repairing pipes, installing gate hardware, repairing picnic tables, painting a door, *etc.* For those special projects, I would typically spend the previous week making up kits for each with a bag loaded with the parts, tools, diagrams, and notes about what was to be done. That method greatly cut the time to explain, making it easier to manage 30 to 35 people waiting to be assigned a task or asking for guidance.

During my first season, the Board asked for more shade for poolside chaises and encouraged me to build something during the next year's work sessions. I looked at shade structures at other clubs, making notes of good and bad features and consulted an ephemeris to determine sun angles at various times over the day and dates of the Club's season. During the following winter I designed a wood structure that was lightweight and not too tall and overbearing. I was averse to using pressure-treated wood because of its chemical content and uncertain repute with respect to health, especially next to the solvent effects of a pool. Therefore, I chose to build the structure of 4x4 redwood posts top-braced with 2x4s and covered with removable slat panels to avoid having to build it to withstand snow loads. Each fall, the panels were removed and stored indoors leaving a skeletal frame to easily survive the snow. They were quickly secured in place with plastic cable ties the next spring.

In my last year as B&G manager, I oversaw the rebuilding of four tennis courts, a new education for me. After the third year, I took a year off, then got elected as president for two more years.

We, and especially our youngest son, thoroughly enjoyed our summers at WSTC for a dozen years but our use of it faded soon after he went to college.

Road Construction Committee – In 1989, I was asked to consider serving on the committee overseeing reconstruction of a collection of badly-failing roads. My appointment began a 12-year run of serving the Town of Wayland.

By the time I was appointed, the RCC had already repaved several short roads with few challenges. Those roads had, for the most part, been ground up, graded, and repaved. But most of the roads at issue were long, had drainage and subsurface problems, had rights of way issues, and needed to be engineered. An engineering firm had nearly finished surveying these roads and was beginning the design phase. Our committee of seven members met weekly with the engineers and reviewed the problems of the week. Our part was to assure the protection of trees, minimize hardship on abutters, and to generally maintain the scenic character of the roads.

Since many of these issues affected abutters' front yards, their trees, their stone walls, their 100-year old yews, *etc.*, one of the first things I did was to walk these roads, take pictures every 100 feet (using the station marks painted on the road), take extra pictures at intersections, and GBC-bind them into booklets for each road group being engineered. It turned out to be very useful. Many times as questions came up, from either the engineer's technical perspective or abutters' aesthetic perspective, we were able to pick up the book, turn to the relevant pictures, and say, "Oh, *that* wall" or "*Which* tree?"

I served on the RCC from September 1989 until September 1991, one year as chairman, and enjoyed every moment.

Finance Committee – After two years on the RCC, an opening occurred on the Finance Committee. By that time, I was known to the Board of Selectmen; they respected me, and chose to appoint me to the Finance Committee.

During my third and fourth years, I was chairman. The Town Accountant and Finance Director, was at all our meetings and was usually the source of our money-information and issues. The FinCom prepared the budget each year and reviewed all of the articles submitted for action at Town Meeting.

Naturally, I could never volunteer for anything without undertaking some kind of engineering or fix-it project. You wouldn't expect the FinCom to be fruit for such undertakings, but it had some. And I found them by asking to review any articles that had meaty technical aspects associated with them: the

Town's new computing system (I also served on a briefly-formed committee for that major upgrade); refurbishment of the Town Clock (I toured the clock tower); trails and bridges through Town land; the Water Department's many projects for new pipes, pumps, wells, and monitoring systems; software for the Town Surveyor; saving of the swimming pool; highway projects; other similar issues; and the two described below.

FIX-IT PROJECTS – One year, the Town Clerk's office came to us with a quoted proposal for about $30,000 to install a new door on the vault, an all-concrete-block room for files and historical records. The vault door was a foot thick with a six-number combination lock. We didn't have an extra $30 k lying around, so we challenged the need. The reason was that the lock didn't meet current code of having an interior release to allow an entrapped person to escape; besides it was very hard to pull open once unlocked. I offered to take a look at it before we decided.

I discovered that the door originally did have an escape mechanism—the release button and the actuating lever were still there but a recent replacement of the lock did not include a connection between lock and button. I arranged to meet up with the Diebold repairman and ask about options. There was only one combination lock model that would fit that door and it had no release feature. He left a new lock with me. I pondered it and eventually worked out a design for a simple modification that interfaced with the door's unbolting system and with the red release button. I had a shop modify the bolt and make two new parts. Diebold returned and for a few hundred dollars installed the modified lock. Presto, the Town had a vault that met code.

But what about the hard-to-open problem? Apparently, the door hinges had worn enough over the years to drop the door tightly against the threshold. Someone had already ground the threshold downward, but the grinding did not extend fully into a deep corner. I sharpened my cold chisel, spent 20 minutes scraping that last little bit away with hammer and chisel, and the five-ton door opened like velvet. Thirty thousand dollars saved.

Another project affected the historical Gate House. Several years in a row the Conservation Committee asked us for money to replace the roof of a small building on the dam of a Town-owned lake. The original purpose of the gate house was to enclose a water-driven pump used to generate water pressure for putting out fires in the Cochituate section of town. The building also included a large drainage gate valve: hence the term *gate house*. Several of us were taken to the building where we found a beautiful stone building with a decorative slate roof and wall areas styled in *opus reticulatum*. Although the gate house was effectively landlocked by surrounding private property and therefore inaccessible to the public except by canoe, we eventually decided that its uniqueness deserved preservation. We approved the requested $35,000.

But the lowest bid came in at almost $150,000 primarily because the badly rotting roof would require extensive scaffolding or a crane to remove while preserving the unusual and decorative slate. Another dead project. But I had an idea, consulted with a young clever carpenter I knew, and together we came up with an inexpensive way to take down the old roof safely and begin construction of the new one. I prepared a full set of drawings for him to bid from. He told me his price, and I advised him to double it to a still-low price and submit the bid to the Town. He got the contract, and the Town got a new roof well-under budget. He finished the roof with asphalt shingles, but he used the same colors of gray and red and in some cases cut each shingle to match the original patterns of slate used in that area of the roof.

I served on the FinCom for ten years from 1991 into 2001. It was a rewarding experience.

THE GTE YEARS
December 26, 1978 to September 29, 1997 – Ages 38 to 57

I started work at GTE on Boxing Day in 1978. I intentionally started the new job before the end of the year as a result of the inadvertent but beneficial beginning of my Polaroid career two weeks before the New Year. But it turned out to be an unnerving experience. Since GTE did not allow unused vacation time to carry over to next year, most of the staff was away during that week between Christmas and New Year's Day. Moreover, this being the '70s, the remnants of the energy-saving mentality following the 1974 energy crisis was still with us, and GTE had no lights in the hallways except for an occasional safety light every 100 feet. The place was dismal: dark offices, dark hallways, dark laboratories. It felt abandoned. None of my future associates were around and only a temporary secretary was able to direct me to my office. I thought I had joined a company in the throes of going under.

But the next week finally arrived and so did the bedlam of an active technical center. The place was GTE Laboratories Incorporated.[13] The Labs were considered the central technical resource for the entire corporation and derived their funds from a "contribution" levied on each of the divisions. Most divisions also had their own research and engineering centers to carry out their daily technical needs. I was a Member of Technical Staff (MTS).

My new supervisor, Vince Oxley, hired me into the Electrical Equipment Group to oversee the design of a **modular circuit breaker** that had begun some months earlier for the benefit of the Electrical Equipment Division of Sylvania, also a division of GTE. Sylvania had acquired several companies that built, among other things, ordinary household circuit breakers. The result was that Sylvania was then manufacturing five distinct configurations of circuit breakers that fit into five distinct breaker panel designs. Thus, none of them could be discontinued, and none of them was being manufactured in adequate volume to warrant automating their assembly. The plan had been for the central research Labs to design a circuit breaker module that was common to all five breaker housings and common to all ratings from 15 to 100 amperes that could be completely assembled and tested by machine, then dropped into any of the housings for sale. Great idea. Not very practical. Anyway, it had become my job to make this work.

> **BREAKER DESIGN** – The extant design team comprised a contract engineer with plenty of experience but not much imagination and two part time drafters. The machine shop was well-equipped with every needed resource including numerical-controlled milling machines. The design status was a nearly complete set of drawings that depicted a design copied verbatim from a 4-times size plywood and Plexiglas prototype contrived and hand-made by a nonresident consultant some

---

[13] When I joined, the company was General Telephone and Electronics Corporation, generally known as GTE. In 1982, the official name was changed to GTE Corporation.

months earlier. The design itself was a compromise because of its need to fit within both a long skinny breaker and a short thick breaker. The module was compact and crowded. The shop was struggling with converting highly complex shapes described by the old-style drafting methods and was milling the urea-formaldehyde cases from toilet seats made of the same material. This project needed more than mere leadership.

One of my first actions was to meet with the shop manager and a key machinist to discuss their needs for making parts efficiently. Next, I set myself up as chief designer and drafter to virtually start the design again, but using a dimensionally-stable Mylar film to control the overall geometry drawn at four times size at a drafting board. Though some details were done by others, I did the complex ones using ordinate dimensions that could be quickly and accurately input to the numerical-controlled milling machines. My work also included designing the tools for forming two of the metal parts with drawn or formed-up edges and features. Once this phase of the project was moving, I undertook the job of how to simply mold the housings using proper materials instead of machining the parts out of toilet seats. The shop made a 3-piece aluminum mold that could be charged with urea formaldehyde resin and put manually into a heated laboratory press for squeezing the material while hot to about 2,000 psi. After a few minutes, it was cooled, opened, and the "case" or "cover" removed by tapping evenly on about six ejector pins. They took about 20 minutes each, and I had patience enough for only about three parts per day, but my technician was able to keep going.

When the semiannual project review rolled around, we were making parts in small quantities, assembling a few breakers, and initiating test protocols. One of the attendees at the project review was a hard-driving VP at Sylvania headquarters who liked to get things rolling. Before the meeting had even taken its break for morning coffee, the VP had called his favorite outside shop, asked the two owners to come immediately to the Labs, and by lunchtime I was talking to them about making 1,000 parts for us so that we could get serious about progress. That shop turned out to need a lot of help from us and was unable to finish the case molds or to mold the cases. The mold had to be finished by one of Sylvania's regular vendors. We spent more time refining "volume" prototyping techniques than refining an effective working design.

Ultimately, we built and tested several dozen 15, 20, and 30 ampere breakers, but the compact design could not withstand the load required of the larger capacity breakers. The flexible cables broke before the required 10,000 switch cycles, the magnetic switch function with its too-short bimetal could not trip fast enough at 120 or more amperes, and the 10,000 ampere tests blew the breaker cases to smithereens. Fortunately, by this time I had taken on enough other projects to keep me busy while changes to the breaker design were being made in the many attempts to eliminate the flaws. It was hard to envision how the 100 ampere breaker could ever survive if the 30 ampere breaker was doing so badly, but my supervisors were ever optimistic.

The Electrical Equipment Group included about 20 engineers, scientists, and technicians working on a variety of electrical projects—the science of electrical arcs, electronic circuit breakers, computerized numerical modeling, heavy-duty motor speed control, running a first-class high-energy electrical testing lab, and supporting the modular breaker project. Soon, I was helping some of my colleagues with their projects in addition to my first priority, the modular breaker. It was good work. It had the variety found at ECA and the sophistication found at Polaroid. The facilities were clean, needs were usually satisfied, and typically there were four to eight people working for me on the modular breaker at any one time. The modular breaker work did not result in any patents, but design work on other breakers did. (See US Pat. Nos. 4,464,641 and back cover center right; 4,472,696; 4,472,701; and 4,491,814.)

On two occasions in my first two years, I flew to Puerto Rico for a technical meeting, a management meeting, and to visit three factories.

• • •

In the early 1980s, GTE offered the **newly enacted 401(k)** to its employees. The Labs conducted informative seminars to which everyone was invited. As was my habit to foster long-term fiscal advantages for myself and family, I signed up—and to the max, which at the time was "contribute 6% and the company will match 3%." Later the maximum contribution was increased, and I always pushed the limit even when it hurt in the paycheck.

GTE also had a **stock purchase plan** through which stock could be purchased at a 15% discount. The plan was so generous that to purchase all the shares offered would have made a major dent in ones paycheck. It was also not a guaranteed win—only a discount on a publicly traded stock that might rise and fall with the market. I did however cautiously enroll each year.

Through my analysis and judgment, both of these opportunities were good channels for working toward my objective of enjoying a comfortable life in retirement. And that's what it eventually did—sustain a decent income even after leaving the workplace.

• • •

I also did some "consulting" for Chris Peek, one of my bosses of years earlier at ECA, who was now head of another department. I did some mechanical design for three of his products: a **stand-alone programmable controller** for which its electronic designer and I designed the structure;[14] a **modular control system** for which I devised a zero-cost but very effective captive mounting screw, designed a mold, and molded nylon housings; and the **redesign of a solid-state relay** that couldn't be molded as originally designed. Also, Peek got involved in helping the University of Chicago build about fifty **high-temperature solar collector tubes** of novel design. A graduate student spent about six months at the Labs overseeing their construction, which required some extensive work by our glass shop and elaborate fixtures made by our machine shop. The grad student often came to me for advice and design help, one feature of which resulted in my twentieth patent (see US Pat. No. 4,440,154 and back cover center).

Toward the end of 1982, GTE decided to sell Sylvania's Electrical Equipment Division. By that time it had become obvious to all that the modular breaker design would never fly, so with the impending sale of the recipient division, all work on the breaker ceased. This sale would affect the entire department. Many of my colleagues were looking for new jobs within GTE, and for a while, my only remaining assignment was to write a summary report on the status of the circuit breaker. This was the first of many reports I eventually authored at GTE, a list of which is found in *Appendix A*.

• • •

---

[14] The *Intelligent Industrial Relay*, or I^{2}R, was later featured on the cover of *Control Engineering*, May 1982.

Happily for me, an ideal opportunity came up. The Director of GTE Labs had "purchased" four **Get Away Specials from NASA** with hopes that they might be useful for some future zero-gravity research aboard the Space Shuttle. The Get Away Special (GAS) provided a 20-inch diameter by 28-inch high volume, a 200-lb weight limit, and 3 throws of a switch by an astronaut. The lighting division was proposing to test a **metal halide arc lamp in the absence of gravity** using one of these payloads. In January 1983, I was asked if I would like to build the payload. I was giddy with anticipation of making this work.

That project turned out to have a lot of surprises. The first surprise was that NASA did not supply any power for GAS experiments, so we needed to include *battery power* to operate a 175 watt lamp with alternating current. Another surprise was the amount of testing and documentation required to demonstrate that our payload enclosed in their container would not jeopardize the Shuttle mission. There were four people working on this project together, and except for me, all were plasma physicists more interested in arcs than in hardware—it was up to me to dream up most of the actual hardware. For example, they needed to study the image through three different narrow-band filters to separately observe the mercury, sodium, and scandium, but I had to figure out how to do that and where to get the filters. Later, I had the help of Glenn Duchene, an electromechanical technician who worked on the project full time in the later months and was of immense help.

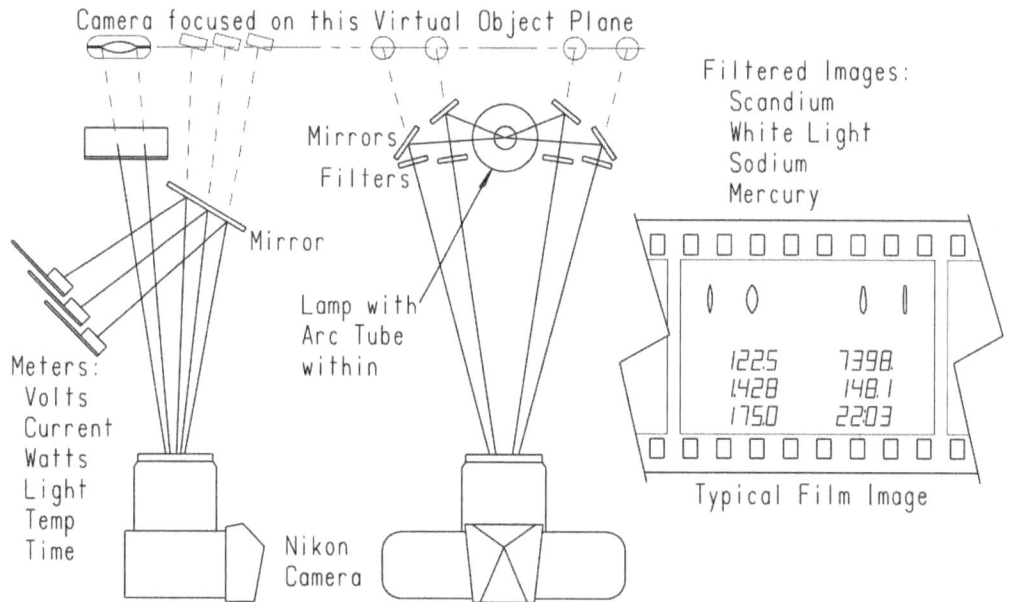

The optical system for the lamp-experiment payload. At the bottom are side and top views of the camera and the ray-tracing of reflected light that combine four images of the arc tube and six meter faces, all on the same virtual object plane for sharp focus of the images. At right, a diagram of the resulting negative.

I threw my heart into the project and came through with a lot of successes. One was an optical system using five mirrors for recording four differently-filtered lamp images simultaneously with six digital meter faces to record voltage, current, watts, temperature, light output, and time on the same frame of film—and all in the same focal plane for sharpness. Another was finding the best energy density in ordinary alkaline cells, especially the Duracell brand. Final choice: an F-cell, 50% longer than a D-cell combined into 19 hexagonally-arranged groups of 7 hexagonally-packed cells that fit into the circular GAS container with about $1/2$ inch to spare. Another was that we had sufficient power that we could include three redundant, independent experiments for greater reliability.

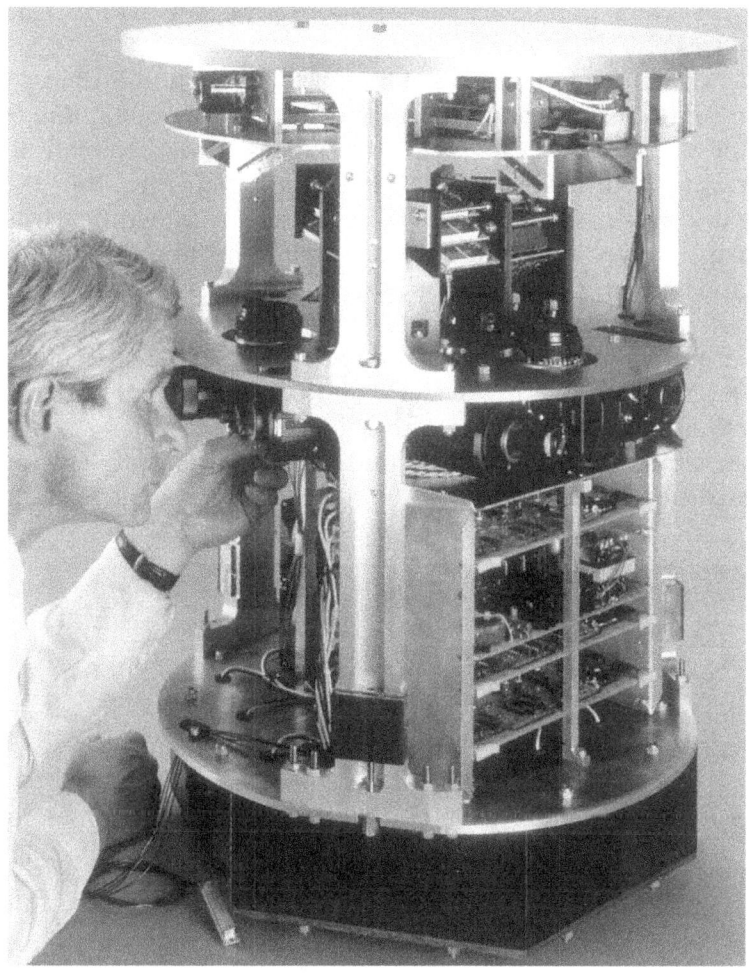

The first Get Away Special experiment designed and managed by me in 1983. From top: NASA cover, two of the three metal halide arc lamps visible, digital displays, cameras, electronic racks, and batteries within the black hexagonal box.

Despite having mostly part-time support, I implemented or oversaw many diverse aspects of the project: the design, numerical modeling, building, and vibration-testing of the structure; selecting and verifying the functionality of three Nikon cameras with long rolls of film for 256 pictures each; hiring a consultant to design the solid-state lamp ballast; lay out, prototype, and replicate three ballast circuit boards for driving these lamps in both AC and DC modes; defining the hard-wired circuit for controlling the 30-minute experiment comprising the 60 Hz AC power driver, the timing of 256 pictures at 7-second intervals, a 15-minute AC period, four positive DC periods, and four negative DC periods; finding a vendor for coating high-temperature multilayer filters to separately transmit light from the mercury, sodium, and scandium species of these complex metal halide lamps; and the endless work with NASA and their required documentation. In the middle of this project, I took the entire month of July off to take the family on a Winnebago journey from Denver to Denver via Yellowstone, Yosemite, San Francisco, Los Angeles, the Grand Canyon, and many other places along the way.

Glenn, my technician quickly evolved into a full-time, high-talent assistant who continued to work with me for much of my remaining career with GTE. With all of us working together, our payload was complete in only ten months. It was boarded onto Sylvania's corporate jet the Friday after Thanksgiving and taken to Kennedy Space Center for integration into the GAS payload system. Only Glenn and I were traveling on this luxurious corporate jet, and we landed at the Titusville Airport, spent three or four days at Patrick Air Force Base, and watched a Shuttle launch. **Our payload, G-051, was launched on February 3, 1984 aboard Challenger, STS-11** (officially STS-41B in a short-lived goofy numbering system) and returned to us about four weeks later. My wife and I attended the launch. All three systems worked nearly flawlessly except that the battery drain and their unexpectedly low temperature during the third experiment combined to cut that experiment short by about 8 minutes.

1996 US postage stamp of STS-11, aka STS-41B (incorrectly labeled STS-7). Our payload is the middle of the three unevenly spaced dark dots along the starboard edge of the cargo bay.

But all good things must come to an end, and so it was with the Shuttle project. Between delivery and return of the payload in March 1984, I managed to latch onto some miscellaneous projects to keep me occupied. Once the payload was returned, we were again busy examining the lamps, developing the film, plotting data, building a jaunty-angled rotating stand for displaying the payload, writing a summary of the project for the Get Away Special Symposium at Goddard Space Flight Center, jointly authoring several scientific papers, and making presentations illustrated with slides at several Sylvania sites.

Years later, I discovered that a $3.00 postage stamp was issued in 1996 that pictures the Shuttle with our payload hovering over Earth. (The stamp is marked "Challenger STS-7" but when comparing the stamp's image with those of STS-7 and STS-11, it is obvious that the stamp matches the layout of the latter, our launch, in every respect.) The photograph was taken by Bruce McCandless from 50 to 60 meters on February 7 during his history-making untethered extra vehicular activity (EVA).

But the tailings of the Shuttle project were not enough to keep me busy. One of my projects during 1984 was consulting two or three days a week for Government Systems in Westborough where I was designing a **high-energy surge suppressor for Minuteman Missile** control centers at various sites around the country. These control centers were "live-in" bunkers buried well below ground level so that they would survive a nuclear attack. They were sizable and required a lot of 240 volt power. But a nuclear blast would induce an incredible level of current into the power grid, much greater than a mere lightning bolt. So the power line into these bunkers required a surge arrestor capable of shorting out an impulse of 600,000 amperes to isolate and protect the equipment so that it could continue to operate on battery power—a giant version of the surge suppressors we plug our computers into.

As with the household power-strip kind, the Minuteman surge arrestors would use a metal oxide varistor (MOV), which is "open" at normal voltages, but shorts to ground whenever the voltage rises above about 150% of normal. My part was to design the hardware for supporting and clamping the MOVs, which were three inches in diameter and thin, rendering them quite delicate. I performed some numerical modeling of the current distribution at the pressure plate-to-MOV interface to determine the needed pressure. It also needed a very compact pressure means, and everything needed to be insulated for 240 V. I finished the design on schedule and later saw the results of the 600 kA testing. Many components were melted together, as expected, but the real surprise to me was what counter currents can do to the hardware as the cables are repelled by the enormous magnetism. The 2-inch steel conduit was stretched out into an oval as the two cables were pushed apart, and the $3/8$ inch thick walls of the equipment box were blown out or distorted badly. I was proud to have successfully executed this design, but sorry that GTE did not get the follow-on contract to build the equipment. The low bidder got the action.

Another fill-in project in the 1984-85 period was with a small group trying to develop **thermally-efficient packages for electronic chips**. At the time, GTE was developing new solid-state telephone switches that required very large-scale integrated (VLSI) circuits that produced a lot of heat. Our work was to improve heat extraction. Glenn and I demonstrated a high-capacity cooling system that exploited both the expansion and evaporation effects of liquefied carbon dioxide (see US Pat. No. 4,838,041). The project didn't go very far, but I had fun running around the country to various symposiums and factories: Washington, Baltimore, New Orleans, Vancouver, and GTE's Chicago factory with 7,000 employees.

• • •

In 1982 I was asked to help establish the GTE **Engineering Associates Development Program**, which was to be a 20-month development period for Master of Science graduates. I helped write a description and guidelines for its implementation and did some recruiting. There were three rotations: engineering, production, and managerial. Subsequently I supervised two participants for their six months at the Labs.

In the 1990s, I also participated in a program for **United Negro College Fund students** for three summers. These two- then three-week tours barely gave enough time to accomplish anything, but I tried very hard to plan and have the necessary resources for them to finish a meaningful if small project. One did a pretty good job programming a simple monitor for a battery; another was mostly interested in her social life and accomplished little; and the last did a spectacular job of programming a Tattletale computer to interface with a telephone modem, make and receive calls, and transfer data. I used some of his work product for the Back Watch instrument reported below.

• • •

In mid-1989, I was promoted from Principal Engineer to Staff Engineer, a position that was equivalent to a manager, but on a technical track. I reported to one of five directors instead of a manager, got carpet installed in my office and all new furniture of my choice. Of more interest to me, it put me in a position of being able to support anyone and any project at the Labs. This led to a number of interesting projects. One required me to commute to Salem to design **sputtering equipment** to improve the coating of a novel, flat screen video display formed on a sheet of clear glass. Another was working with two colleagues to design and mold an **angstrom-accuracy lens** for the tip of an optical fiber (see US Pat. No. 4,953,938).[15] Four of us worked together on an **optical fiber crossconnect switch** (see US Pat. No. 4,955,686). I designed a **motorized probe** to explore the plasma within a fluorescent lamp. During the rage of high-temperature superconducting materials, I worked with a materials scientist, to conceptually design a **formed-to-shape superconductive magnetic coil** using one of the coated types of superconductors (see US Pat. No. 5,173,678). I helped a colleague with both a **microwave-excited video display** suitable for large JumboTron-like applications such as seen at Fenway Park; and a **microwave-powered automotive headlamp** (see US Pat. No. 5,299,100 and front cover center right). Another materials scientist was designing a **dielectric resonator** to power cell phone antennas. The material worked great, but it needed to be suspended in the center of a metal box without touching anything. He asked me for help, and a glass engineer and I came up with a solution (see US Pat. No. 5,347,246).

---

15 The process we used for making the mold insert for this high-tolerance lens surface was diamond machining of hard-plated nickel, the same process used for making the tiny, highly accurate aspheric lenses needed for smart phone cameras.

Also, I was periodically called upon to visit a factory here or there to consult on a product, hardware, or assembly issue. I went to factories in Puerto Rico and visited the huge, highly-automated factory for telephone equipment in Northlake, Illinois. For the Sylvania division, I visited the glass-processing factories in Rhode Island and New Hampshire, and consulted at several processing and production factories in Massachusetts, New Hampshire, and Maine.

In the spring of 1984, I made a two-stop trip to Boling Air Force Base and the New York Times. The first stop was with three others to make a pitch to NASA and the Air Force to build another GAS **Space Shuttle experiment for growing gallium arsenide (GaAs) crystals** without buoyancy-driven convection in the "melt crucible." The second stop was with my director for an interview with John Noble Wilford, the science editor, about our metal halide lamp experiment. It was a nice, but brief article that included one of the photographs made using the multimirror structure that I invented. (*The New York Times*, Tuesday, June 19, 1984, p C3.)

The NASA-Air Force connection ultimately resulted in my major project from 1985 into 1992. Our first payload had taken 10 months, why did this one take so long? Mostly to fill the time available, *a la* Parkinson's Law. It took about 18 months after our first presentation before NASA and the Air Force came through with a contract. During the wait period, I contributed to some proposal-writing and gave the crystallographer, Jim Kafalas, some hardware support. By the time the contract arrived, I had already designed most of the payload structure, so it didn't take long before parts were being made in the shop. Also, on January 28, 1986, the first shuttle disaster occurred, which delayed subsequent launches. That failure was during the launch of Challenger, the same shuttle that had carried our lamp payload two years earlier.

During the early stages of this program I researched the status of **computer-aided design software** and ultimately purchased the *Cadkey* design software. Until that time, I had executed all my design work on a drafting board. It took me about a year to evolve from designing on paper to designing on the monitor. The most satisfying result was that all the parts fit together perfectly. Since the computer included the original fits and relationships, the parts got dimensioned accurately. My experience with designing on paper was that although I worked quite accurately, among dozens of interacting parts and thousands of dimensions, there would inevitably be at least one feature incorrectly dimensioned, or one stock thickness omitted, or one hole marked with the wrong diameter. The computer-based design process had the capacity of transferring fits from one part to another, thereby minimizing fit problems. It didn't, however, make the designs any better.

> **DEVELOPMENT OF A SHUTTLE PAYLOAD** – Despite the shuttle disaster, we plowed ahead on the project. The plan in its simplest form was to determine if an already-grown crystal of gallium arsenide, GaAs, could be regrown without buoyancy-driven convection and result in fewer defects in the crystal lattice. Earth-based experiments had already established that the turbulent pot of molten material

was at least partially responsible for the typically large quantity of defects in GaAs crystal formation. Was it possible that the lattice defects could be greatly reduced or eliminated by growing the crystals without gravity? Kafalas had demonstrated in the laboratory that a previously-grown crystal could be machined to fit closely in a quartz test tube, the upper 80% melted at 1,238°C, then cooled very slowly to regrow a single crystal from the unmelted "seed" crystal at the bottom.

It was clear from the beginning that this experiment, unlike the lamp experiment, would need a **low-power computer** to control the complex process of growing a crystal with its requirement of fractional degree control of temperature over many hours. I found a computer that could be powered many days with a small battery, but it needed to be programmed in assembly language, which was not for me. A programmer and his technician were eventually assigned to the project. They spent well over a year, writing thousands of lines of code and hadn't yet demonstrated success of a single control algorithm. One day, the programmer came to my office and told me that he was taking a job out of state.

I went into major panic mode. This was spring 1987 and we had planned this project for nearly three years, worked earnestly on it for a year and a half, and the control system was back to square one. No other programmers were available. I followed up on a lead about a low-power "Tattletale Data Logger" made by Onset Computer Corp. on Cape Cod. The inventor and owner told me by phone that Tattletales could do far more than just collect data. On a Friday, I visited their company located in a barn. The owner showed us their new Tattletale IV and within a few minutes demonstrated one of our needed algorithms using BASIC. Highly impressed, I bought the instruction manual with pocket cash and by Monday morning had learned enough TT-BASIC to write much of the furnace-control algorithm. First thing that morning I ordered a Tattletale IV, got myself outfitted with a terminal emulator and text editor, and by the end of the week was running and testing my program with simulated inputs.

Next we coupled it to the crystal-growing furnace. Glenn already had the many signal conditioning components incorporated into the previous computer system and was able to quickly jumper enough wires and terminals so that we actually controlled a furnace with the low-power computer and our own software another week later. This particular computer was running interpreted BASIC and required numbered lines. One immense advantage of such a primitive configuration was that with a pair of keystrokes I could momentarily stop the program, rewrite a faulty line of code or type a new but intermediately-numbered line of code directly into the computer and have it resume where it left off, but with the corrected software in play. This was very useful in the early stages of developing this software with a typical run time of seven to eight hours coupled with an expensive crystal ingot in a 1,200° oven.

Generally, semiconductor substrates such as silicon or gallium arsenide are grown by touching a seed crystal to the surface of a molten pool of the material then slowly lifting the seed upward such that the liquid freezes into a uniform crystal that is attached to and copies the seed. This method is known as Czochralski growth and the result is called a boule. Our experiment was configured to start with a previously-grown boule machined to fit snuggly in a sealed quartz glass ampoule, melt 90% of it, leaving the cool end to serve as the seed, and slowly cool the melted portion to freeze, starting from the seed and progressing along the boule.

For our experiment, we moved the liquid-solid interface along the boule not by moving the heater, but by slowly lowering the temperature of a heater configured to have a slightly tilted temperature profile. It was also my job to design that critical heater using platinum wire. I used a variant of NASTRAN, the finite element program developed by NASA for the Apollo mission. Many run cycles were needed to optimize the simulated profile of the wire windings. The actual heater reproduced the simulation quite well.

Making the heater entailed wrapping the wire around an alumina tube with a variable helix. The technique I developed was to trace the variable helix on the tube in pencil from a template, wind the platinum over the pencil mark, periodically apply patches of masking tape to keep the wire stable, and encapsulate it with a paste-alumina section by section so that the tape could be removed before finishing. The built-up and dried paste was then scraped down to a uniform cylinder, and the composite was then fired at high temperature to fuse the tube, wire, and paste into a robust unit.

Quartz, *per se*, is a crystal of silicon and oxygen. The crystal can be melted and fused to become a glass-like material called "fused quartz," but often simply called "quartz." Because of its rigidity at high temperature, it was needed for the enclosing ampoule during growth. Without gravity, the liquid would break up and float around within the ampoule. Moreover, GaAs, like water, increases in volume by about 6% as it freezes, so it would not be possible to shape the ampoule to hug the GaAs on all sides. We needed a plunger and a spring to keep the liquid in place. But what kind of spring can withstand temperatures of 1,238°C?

One day, I visited a scientist friend just to be sociable, happened to see a broken and shredded pot-shaped vessel, and asked him what it was as I picked up one of the smaller pieces. He told me it was a high-temperature crucible made by one of GTE's divisions and the material was boron nitride. I noticed that the piece, less than a millimeter thick and from the curved wall of the crucible, was like a spring and wondered if we could use it for the **ampoule's high-temperature spring**. Eventually, I was able to find more scrap, cut it into little squares, and stack them up with their slight curvatures back to back, thereby forming the stack into a spring with half an inch of stroke. I devised a graphite piston to cradle the spring stack—problem solved.

Because this was a government contract, I as **Payload Manager** was responsible for writing monthly reports, holding periodic progress reviews at both GTE Labs and NASA Lewis Research Center in Cleveland, developing the Payload Accommodations Requirement, meeting with the Technical Manager at NASA Goddard Space Flight Center, providing Hazard Reports, *etc*. NASA coined the catchy name *GaAs GAS*, referring to both the crystal and the shuttle flight program. A few years into this project, another materials scientist, Dave Matthiesen, joined the team. A year after that, Kafalas retired. Thus in the final period, Matthiesen was the crystallographer.

During a few of the earlier tests, the stabilization temperature was so inaccurately attained that either the entire crystal melted or only a small portion melted. We needed to come close to $1/2$ an inch of seed and $3^{1}/_{2}$ inches of melt. My proposed and successful solution was to place a thermocouple within a recessed well in the base of the ampoule, and write a control subroutine that used that temperature, its error-value, and its integrated-error-value to gently approach the target of 1,234°C, a few degrees below the 1,238°C melting point.

The batteries for this project were identical to those used years earlier for the lamp experiment. A three-axis accelerometer module was included. All the control programs, including accessory programs for baking-out the furnace, recording payload temperatures, controlling and logging the accelerometers, and offloading data were written and maintained by me. I developed checklists to assure that preparation for final flight went smoothly and accurately.

A plane from NASA Lewis in Cleveland took us and the payload down to Kennedy Space Center. We landed on that enormous three-mile landing strip for the Shuttle. Setup at the integration site on Patrick Air Force Base went smoothly as the three basic components—the furnace assembly, and two battery boxes—had already been previously assembled and tested. We verified that all systems were properly connected and watched as the NASA techs placed the payload in the flight container, sealed and lockwired the cover. They purged it with nitrogen at atmospheric pressure.

The first launch, as payload G-052, was on June 5, 1991 aboard Columbia, STS-40. It returned to Edwards AFB in California on June 14. Because of the atypical cross-country return of the Shuttle to Kennedy, our pick-up of the payload was delayed several weeks.

Upon examination, we quickly determined that the first crystal experiment fully cycled but that the second had only heated to about 70% of the melting temperature. The batteries had become so cold after the first five days in orbit, despite the heating from the first experiment that they ran short of energy. That second furnace was put aside and reinstalled for a later launch.

The second launch, as payload G-229, was aboard Atlantis flying as STS-45 on March 24, 1992 and landed at Kennedy on April 2.

I was never involved in the crystallographic examination of the space-grown GaAs except for writing up the specification for dicing it up into samples. By this time, Kafalas, the inspiring force behind the program, had retired, the materials division at GTE Labs was in process of dissolving, and Matthiesen was moving out to a new job at NASA Lewis in Cleveland, taking all his crystals with him. If they were ever studied, I never heard about the results.

• • •

Incidentally, we made a good enough impression on NASA that they contracted us to design and build a prototype of a low-budget **Versatile Furnace** that could be easily configured for a variety of furnace-needing experiments. It was to be operable while installed in a GAS container.

> **SAVING A MILLION DOLLARS PER DAY** – Matthiesen used the Versatile Furnace prototype to verify our low-budget modeling of one of his projects. He had a 1-inch diameter by 2-foot long tube containing a crystal growth ampoule, thermocouples, insulation, *etc.* The tube needed to be sealed on earth at room temperature months before processing in space. He needed to know what the internal pressure at room temperature should be if the final pressure was to be 32 psia when being processed at high temperature. He had an approximate temperature profile ranging from 45°C at the two ends to 1,240°C at the center. A team at NASA Huntsville had given him a quote of $1.5 million and six months to model it. His project couldn't afford that. I overheard him grumbling about his dilemma and suggested we should **model it with a spreadsheet**.
>
> He laughed! Computer-based spreadsheets were fairly new, and this was the first engineering exercise he or I ever did with a spreadsheet—it certainly wasn't my last. We arbitrarily divided the tube into $1/4$-inch increments, assigning each increment to a row in the spreadsheet, ending up with 96 rows. For each row (and therefore each increment), we would solve the gas equation, $PV=nRT$—absolute pressure times volume equals the number of moles times a constant times the absolute temperature. In fact, we would solve it twice: once at the high furnace temperature and once at room temperature. (It pays to have remembered an equation from high school chemistry.)
>
> For the high temperature case, pressure would be uniform at 32 psia throughout all 96 rows. The volume of the gas inside had to be manually calculated for each increment depending on the hardware installed in that increment. Some increments were identical. It took several hours to populate that column of the spreadsheet. R comes from a handbook and is the same for all 96 rows. The temperature had to be manually entered for each increment based on Matthiesen's estimated temperature profile. The equation was set up to solve for the number of moles in each increment. So far, the spreadsheet had 96 rows and 5 columns, two of which are the same

throughout, two are manual inputs, and the fifth contains the solution to the equation. The total of the fifth column indicates the total gas to be sealed within the tube.

For the room temperature case, we repeated the above except that the pressure is a guessed number, and the temperature is room temperature for all 96 increments—two easy edits. We again solved for the number of moles and summed that column to display the total. If the guess for the pressure didn't give us the same total number of moles as in the high temperature case, we revised the guess. After several iterations the number of moles matched and Matthiesen had his answer for what pressure to use when sealing the tube.

We did all this in a single day. We "earned" over a million dollars a day for that project. Too bad we didn't get paid. This is what he subsequently verified using the Versatile Furnace. And just think: this mechanical engineer put his high school chemistry training and his mathematical logic to good use for **solving a problem in one day that a team at NASA said would take six months**.

• • •

So, how did I get into **the telephone business?** It was my experience with batteries.

In late 1989, Joe Proud, my director at that time told me that there was a new project request for 1990 from Telops (Telephone Operations in Dallas) that I was ideally suited for because of my success with power efficiency for the GAS payloads. The question was how to provide local power in the event that telephone service to the home was provided over optical fibers? Power would be needed by the optical-to-electrical circuitry and the telephone itself, the glass fiber couldn't carry electrical power, and tradition was that the telephone company provided the power independent of the municipal AC power grid. He assigned the project to me, my first of many for Telops.[16]

I had to become a quick study—all this was new technology for me. I had some knowledge of fibers and connectors from having my office in the same hallway for the past ten years where much of GTE's fiber research was done, but knew nothing about the signaling, specific electronics for the conversion, nor details of plain old telephone (POT) service.

My power study looked into all possible **power sources** that my interviewees and I could think of: solar cells, wind, thermoelectric, primary batteries, fuel cells, twisted pair of copper from the central terminal, grid AC, and household AC. I also visited several companies who were already designing Optical Network Units (ONU) for the fiber-to-phone interface including R-Tech, Raynet, Alcatel, BBT, and AT&T. One of my conclusions after visiting and studying these systems was that with one exception, no one was trying to conserve power. The conventional telephone uses one watt, but only when in

---

16 Over the following years, I flew to Dallas many times to visit the huge facility in Irving, meet with associates, take part in seminars, and present progress reports on our projects. I also flew to Newark for AT&T; Los Angeles for a telephone class at Cerritos; San Jose for R-Tech and Raynet; Raleigh for BBT and Alcatel; Chicago, Washington, New Orleans, and Vancouver for GTE Network Systems, workshops at Automatic Electric, various component conferences, and Microtel; Atlanta for MobilCom; San Antonio to present a paper on power for fiber optic systems; Tampa/St. Petersburg for both telephone and video projects; Seattle for meetings at the Everett office; Pittsburgh and Dallas for installing Back Watch monitors in Erie and Plano; Springfield, Mo. for Branson; and other places now forgotten.

use. These fiber-based systems drew from 3 to 56 watts *all the time*—up to 1,000 times as power-hungry as the current service. My proposal was that GTE Labs should demonstrate a more efficient system, which we did in 1992. With two colleagues, Glenn and Carl Buhrer, we were able to **design, build, and demonstrate an ONU** that drew about 1.5 watts and coupled to the central office with a DS-1 (24 multiplexed telephone lines) signal over optical fiber, thus the hardware could be expanded to serve many telephones.

My report on alternative power systems scaled the results to 1 watt, an easily scalable and potentially reasonable power level. My observation was that sources such as solar or wind power would require a several-foot large device mounted on the roof in addition to craft-maintainable backup batteries for phone service at night or on cloudy or windless days. Such an installation would cost thousands. Tapping tiny amounts of power from the power company by contract rather than through a thousand-dollar meter installation was unprecedented and considered unthinkable. Running copper wires parallel to the optical fiber from the central office defeated one of the reasons for introducing fiber in the first place. However, one watt of power borrowed from the telephone customer would be worth about 70 cents per year. So, why not get the power from the customer, and if the customer balks, then refund him the $0.70 each year. (And, in fact, modern fiber-linked phone systems such as FiOS do get their power from the customer—with no complaint!)

Also in 1991-92, I got an intriguing request from Mobile Communications (MobilCom), the cell phone division of GTE. A small company had approached MobilCom with an **amazing battery charger**. The inventor claimed it could charge a cell phone battery in a few minutes instead of the typical hour and do it with less heating and less long-term degradation of performance. MobilCom's question was, "Is he trying to sell us snake oil?" We did a quick test and agreed that it appeared legitimate. MobilCom wanted more assurance before funding this small company. We proposed and built six or eight testers that used the same Tattletale computer we used on the crystal growth project. Each tester was designed to charge and discharge five batteries using either the conventional charger or the amazing charger. Most of the testers were delivered to MobilCom for long-term testing. It was not snake oil—it really worked: by a lot. The secret was that it charged the battery at unusually high current for two seconds, then discharged it at a moderately high current for one second, and included a couple of short rest periods. The hypothesis was that the short bursts of discharge "cleaned away" any chemical polarization that might have occurred during the two seconds of charging.

In the spring of 1993, The Labs got notice of a **battery-failure problem** being reported by the Tampa operations center. By this time I'm the battery expert. I visited the field manager in Tampa who showed me pictures of thousands of failed backup batteries and took me to a warehouse where I saw hundreds more, many of which appeared to have just melted from heat. The seriousness of the problem had come to his attention following a bad ice storm in central Florida that knocked out power for many hours over a large area of their network. Traditionally, the phone system is expected to work even during a power grid failure, but much of their network had lost phone service within minutes instead of hours after the ice storm knocked out power.

GTE's telephone network in the Tampa area was very modern with much of it being served with distributed electronic phone equipment connected with optical fibers. That's great from a performance perspective, but keeping those isolated cabinets powered-up during power outages requires backup batteries. Typically, each of these Digital Loop Carrier (DLC) cabinets, about the size of two refrigerators, serves about 500 telephone lines and includes 16 to 24 batteries similar to a typical car battery but of a deep-charge variety. A power supply converts 208 or 240 VAC to 48 VDC, which is used to power the phone equipment as well as keep the batteries charged. When the AC power fails, the backup batteries should keep the phones working for eight hours—ordinarily enough time for maintenance craft to deliver portable generators. But following the ice storm, many of these units failed in a few minutes.

The field manager's question to GTE Labs was twofold: why did so many batteries fail, and is it possible to predict the capacity of a battery without disconnecting it for performing a discharge test?

Based on my visit to several of these sites, my first thought was that many of the failures were probably caused by excessive temperatures within these sun-baked and power-intensive systems. These specialized deep-charge batteries are rated for 77°F maximum. The actual battery temperatures needed to be measured. We purchased a collection of "Hobo" monitors from Onset Computer and installed them in a number of these remote cabinets. These early battery-powered monitors were about the size of a Fig Newton, and I set them to collect ten temperature readings a day for six months. I took them to St. Petersburg and with a service craft's help installed Hobos in five different DLCs comprising three distinct types (AT&T's popular SLC-5, Northern Telecom, and Siemens 914), and took pictures of each typical installation. Subsequently, I made up kits and sent them to Palm Springs, Wisconsin, and Michigan along with pictures and had local service craft follow the installation instructions and send back photographs of the Hobos for verification (my first experience with a single-use camera). I also made change-out kits that could be mailed with prestamped return envelopes at the 6-month point. The results were astonishing. All nine DLC cabinets were extremely hot in summer, and overly hot in the winter. The highest air temperature recorded was 149°F in Palm Springs at which time one of the batteries was 156°F—79 degrees above the specified limit.

These measurements continued for a full year and were described in three reports: one with 40 days worth of data, one at six months, and a final one displaying a full year of measurements.

In an overlapping experiment, I arranged to have three wells drilled in the ground at GTE, lined with plastic pipe, and outfitted with similar Hobo monitors. The purpose was to determine the efficacy of storing batteries underground where they should be much cooler. Subsequent to this testing, I designed a battery vault made from a large diameter plastic pipe and installable with the same augering machinery that installs utility poles. The vault would be sealed against flooding and have a screw jack to lift out the battery shelves for maintenance. There was not enough interest to warrant building a prototype—"GTE doesn't *make* any equipment: it *buys* equipment."

Over the course of the battery temperature study there was a lot of discussion about how to determine the quality of a battery that is constantly being trickle-charged either by prediction or measurement. In late 1993, I proposed that GTE Labs build a **battery monitor** capable of making lots of field measurements and tests on batteries in service—charging voltage and current, battery temperature, air temperature, open circuit voltage, transients, *etc.*—and periodically disconnect the batteries to measure open circuit voltage or perform a discharge test. The latter would prove actual capacity. The hope was that somewhere in those data would be found an indicator of battery capacity. The project received approval. It also received a lot of skepticism because telephone systems have a positive ground and operate at negative 48 volts making it hard to monitor with ordinary electronics, and the 24 batteries would require a lot of channels and collected data.

This led to perhaps one of the **most fulfilling, but ultimately bittersweet, technical achievements of my life**, and one that was electronic and required complex programming, neither of which typically fall within my field of mechanical engineering.[17]

> **THE BATTERY MONITOR** – I dubbed this monitor **Back Watch** (for *back*up *watch*er). During the design phase, I often called on many acquaintances for help, but most of the circuitry was ultimately of my design. For example, a suggestion for reading voltages of 24 positive-grounded batteries with a negative-grounded computer was to use 24 separate isolation amplifiers. I eventually reasoned that it could be accomplished much less expensively with operational amplifiers, taking advantage of their only bad feature—inversion of the voltage. That was ideal in our case, and ultimately the circuit to monitor all 24 batteries only needed a few resistor arrays, four multiplexers, and two op-amp DIP packages.
>
> Positive 12 V power for the Back Watch came from the monitored batteries and was achieved with a 48 to 12 V DC to DC converter, which was isolated, making the level shift from negative 48 volts to positive 12 volts elementary. Installation on site was made simple by manufacturing wiring harnesses that mimicked those used in the DLC housings, but with the additional features of intermediate voltage-sensing lines, a thermistor for one of every four batteries, and a 9-pin connector at the Back Watch. In the field, we simply unplugged the extant cable from one battery string at a time and attached our special cable—same connectors, same color code.
>
> I selected the Tattletale 5F computer. It was small, had lots of memory, and ran fast using compiled BASIC. (It was a single circuit board 1.2 by 2.0 inches with 40 connector pins for power, analog to digital inputs, digital input/output terminals, *etc.*) A two-line 32-character LCD display was included to show monitoring status and permit local setup. The LCD continuously scrolled through date and time, alarm status, voltages, currents, and temperatures of the six respective battery strings.

---

17 But remember, my engineering *practice* had long included many electrical and electronic systems including an electrical test box invented before I was a teen; the turn indicator system made for the '48 Ford; the inverted pendulum; the electronic flash; the electronic exposure control for a camera for which I got a patent; the electrical switchgear designed and built at GTE; and the electronic systems I specified, monitored, and approved for both of the space payloads.

Following a test, it listed capacity of each battery. When a setup button was pressed, the display offered change options including date and time.

At the time, I also had other projects, so this one advanced in fits and starts. I already had a small lab and arranged for a part-time technician to build a test rack with shelves for 16 backup batteries, a standard telephone system power supply, a typical group of sockets for attaching the battery strings, and a bank of load resistors for discharge testing.

The 8 by 10 inch Back Watch battery monitor. The 32-character display and Tattletale computer are mounted at upper right of circuit board, the telephone modem at lower right, the high-current relays and sockets along the left, and signal conditioning circuitry throughout. One of six cables is shown connected. The cover is partially shown in foreground.

For measuring current, I experimented with measuring voltage at each end of a shunt resistor and calculating the current. Although generally considered a crude method, I found that with a precision resistor of very low resistance (0.01 ohm) it was very accurate. The voltage measured at each end of the resistor passed through a multiplexer (for sequentially reading all six channels) then to the differential inputs of an instrumentation amplifier. The output, proportional to current through the resistor, fed into an op-amp that offset the signal by 2.5 V so that both positive and negative currents could be read by the positive-only 5-volt A/D converter. The instrumentation amplifier, under instruction from the computer, first read the current in its low sensitivity mode, then, if the current was below a certain threshold, it switched to the high amplification mode for a higher-resolution reading. Thus, the circuitry could measure from +25 to –25 amperes with a resolution of 0.0012 amps for the low values.

I found a low-cost, board-mounted 2400 baud modem that would permit the computer's RS-232 output to deliver data to our lab by telephone (all DLCs have a local phone line with a modular jack). It also allowed us to call into the Back Watch if necessary—which was done a few times.

By this time, the circuitry may have been simple in concept, but it had 30 channels of battery voltages, 6 of current, and 7 of temperature, each one of which had at least two scaling resistors, multiplexers, and op-amps in their path. There was a lot of stuff. How could it possibly measure accurately even if using precision resistors? I configured the software to calibrate each input with a calibration value that would be burned into permanent flash memory during an initial calibration routine accomplished with a separate calibrator-program temporarily running in the Tattletale. During calibration, the Back Watch was hooked up to a set of batteries, the actual voltages, currents, and temperatures measured with a precision meter, each reading typed into the Back Watch's microcomputer, and the calibrator then used the readings to calculate calibration values for each of the 40-odd inputs, and burned them into the permanent, nonvolatile memory. Therefore, every unit had a different set of calibration values specific to that unit's hardware. Each Back Watch unit also had a unique serial number burned into it.

Once all the various functional circuits had been designed and proven, I designed the circuit board using Tango, an early, easily-used PC-based design tool. There were over 50 signal lines to the 7 multiplexers, so I chose to keep the trace layout as simple and orderly as possible while letting the computer organize the scramble of readings instead of the other way around. The circuit board ended up 7.4 by 9.8 inches. I designed a two piece housing of aluminum and had it fabricated by an outside vendor who also silkscreened the face with the name and labels for the display and input connections.

As the design phase was nearing completion, the technician tuned up the modem-to-Tattletale interface, meticulously checked my circuit schematics, confirmed the board layout, assembled the first unit, and put it into test using the previously-built rack. Twenty circuit boards were assembled by an outside PC shop and shipped to us for assembly into the housings. Calibration of those nearly-800 channels took days. The finished unit measured 7.8 by 10.0 by 1.4 inches. A computer was set up in the lab to receive data via nighttime phone calls from the field units.

The program used integer arithmetic (no fractional or decimal values), thus 12.043 volts was recorded as 12043 and all data recorded in binary form, *e.g.* 12043 became 00101111 00001011 and would fit into two bytes as shown. Also, it was important to make computations in the correct sequence to preserve the accuracy, *e.g.* while you might expect $3276 \div 5000 \times 18381 = 12043$, in integer arithmetic the first division results in zero, thus the answer would be zero. Therefore, it must be calculated as $3276 \times 18381 \div 5000 = 12043$. (I had to learn "a new math.")

Eleven Back Watch units were installed in three areas of the country: six in Pinellas and Bayou City, Fla. near St. Petersburg; three in Erie, Pa.; and two in Plano, Tex.

FUNCTIONALITY – The program surveyed voltage and current twice each second. If measurements were within normal range, data were recorded into memory only once every 6 minutes. Every recorded reading was taken multiple times and averaged. If everything continued to be normal for the next two hours, the 10 sets of values in that earlier hour were averaged, recorded as a single set for that hour, and all subsequent data were moved back in memory by 9 sets-worth. If the monitored values did not appear normal, *e.g.* a power outage, the 6-minute cycle values for the previous two hours were preserved for detailed review. If a power outage occurred, data were recorded at an even higher rate for subsequent study. Every memory record comprised 100 bytes and included date, time, a status code, ambient temperature, six currents, twenty four voltages, and six battery temperatures.

Once a month, one string at a time was disconnected for an open-circuit voltage test. Late at night during the first full month after installation and every six months thereafter, the program transferred a relay in one of the battery strings to initiate a discharge test of that string through a separate and external resistive load of about 10 amperes. After four days of recharging, the next string was tested. This repeated until all strings had been tested, and if a power outage ever occurred during a test, the test would be aborted. Thus, the telephone equipment should never be low on backup power by more than a small fraction. The discharge data included ampere-hours and watt-hours.

Once a week in the early hours on a staggered schedule based on the serial number, each Back Watch made a phone call to our lab's computer and downloaded the data for the previous week. My technician extracted the data from the lab's computer and pasted it into the respective Excel file for each site. A week's worth of data took up to 10 minutes at 2400 baud. One Erie site had a very noisy phone line and sometimes had to download two or three weeks' worth of data at a time.

The hard part of developing this program wasn't writing the roughly 1100 commands; it wasn't the many subroutines for collecting data, calibrating values, memory storage, memory compression, running the display, making weekly phone calls, downloading data, receiving phone calls from afar, uploading code changes; and it wasn't programming the monthly open circuit tests or the six-month load tests. The truly difficult part was the testing and debugging. Although the program looped twice every second, the only significant functions occurred only once every six minutes, or every hour, or just some hours, or at midnight, or just some days, or once a week, or once a month, even once in six months. Since all coded events were initiated within a second of the top of the minute, I could speed up the program significantly by temporarily inserting an instruction that advanced the clock from second number 2 to second number 59. But that method increased the speed by only a factor of 30—an hour still took 2 minutes to simulate, and a day took nearly an hour. That might be bearable for verifying a working program, but a full day is too long to wait just to learn that a modified one-month subroutine has a missing comma. Therefore, I had to apply additional speed-up, but it had to be done judiciously or there wouldn't be any data collected, or the collected data wouldn't have been compressed, or the program loop might even skip over the one-month subroutine altogether. Eventually, it was all tested, but it took ages.

The bitter part of this sweet program was that after a year of collecting millions of data points on about 250 field-installed batteries, we could not identify a single reliable indicator of battery quality. Every plot of battery capacity versus some other variable—trickle charge current, charging voltage, average summer temperature, open circuit voltage, midstring offset voltage, voltage sag during early discharge, *coup de fouet* performance, battery age, *etc.*—looked like the buckshot pattern from a shotgun. Even the average temperature over a battery's life was, at best, a weak indicator of capacity.

At the end of 1996, we published our final report on the Back Watch project and its findings. The only positive aspect was the successful demonstration that it would be possible to disconnect and perform a discharge test on a regular and staggered basis and report inadequate batteries either by phone or a locally-displayed warning light.

I proposed designing such an instrument, but it was turned down because GTE doesn't *make*—it *buys*. After I retired, I proposed it again as a no-cost ABell Engineering consulting project if they would buy any successful product from ABell. Even that proposal was turned down because ABell was not on their approved vendor list. Sad! I wanted to build that instrument.

• • •

In the last few years at GTE, I got involved with a variety of other telephone-related projects. My experience with power led to many of these opportunities. Some of these projects were to review the powering schemes for ONU test installations proposed by outside vendors. At one of my visits in Branson, Mo., the supplier had proposed using a **parallel coaxial cable to carry the power for the ONU**. By this time I had acquired and customized software to compute power and signal losses in a complex coaxial network. I found that the coaxial cable was so inefficient as a power carrier that the increased coaxial cable cost associated with power distribution alone was greater than if they installed a separate copper cable for the power.

Two of us undertook a study of currently available **pair gain telephone systems** for use in rural areas to provide multiple calls over a single phone line. To save the high cost of a copper-simulating test card for each phone in a fiber-deployed system, I proposed a more efficient test system—a **digital diagnostic system** for distant terminals. (See US Pat. No. 5,937,033 and back cover bottom left.) Telops asked me to evaluate the efficacy of some already-installed **underground battery vaults**. I visited two sites, installed Hobo temperature monitors, and issued a report on the results. And another study focused on **lithium batteries and flywheels** used for backup power.

So many of my projects used or touched on **digital transmission of data** that I began to study several protocols in depth: the RS-232 signal; the DS0 stream; a CODEC's functionality; the most popular scrambling-descrambling algorithm; the T-1 or DS-1 extended super frame; and the high-speed SONET frame and connectivity. I never became a true expert at any of these, but did conduct a couple of seminars.

There was a large trial installation taking place in the Tampa, Fla. area that provided **video and dial tone service via coaxial cable** (future VOIP), which they were implementing in a new "neighborhood node" configuration. Everything was working fine except for what they felt to be an inefficient and expensive power configuration. After several partially successful iterations, I eventually reduced the cost by about $30,000 per node with a plug-in device about the size and function of an automotive fuse. (See US Pat. No. 5,920,802 and front cover, top left for which I drew the figure using Cadkey.)

An off-the-beaten-path project came to me because I was the only mechanical engineer around, but it just happened to be telephone-related. A Pennsylvania division needed an **expert witness in an electrical injury case**. A roofing worker managed to grab a handy nearby cable for support as he was standing up while near the edge of a two-story flat-roofed building. Unfortunately, the cable was a 5,600 volt power line. He caught on fire, was put out by his buddy, but ultimately lost his right hand. He sued the power company for having an electrical cable so close to the edge of a building, and the power company enjoined GTE in the suit because of our alleged part in causing the poles to tilt and the electrical cables to sag so close to the building. These poles belonged to the electric company, GTE was a tenant, and by the tenancy agreement, if GTE were found at least 1% liable, they would bear 50% of the judgment cost.

By the time I became involved, the power company had replaced the leaning poles with taller ones and reinstalled the cables at proper clearance. The offending measurements had been recorded on a drawing that omitted the slope between the two poles. I went to York, Pa. to meet with the attorneys and to visit the site. While on site, I made more measurements, using my digital level and a 10-ft 2x4 purchased locally. With that jury-rig, I was able to gather the missing slope down the slight hill along the building. I studied parts of the electrical code never even seen before, learned about the catenary and the iterative methods needed to calculate its sag, and developed a spreadsheet to calculate the needed dimensions.

What I determined: At most, those big black telephone cables attached to the poles weren't as heavy as they looked and only had a small possible contribution to tilting the poles; the poles had never been properly guyed or braced underground as required by code; and that even if the cables hung without *any* sag whatever they were inadequately spaced from the roof. The cable was only about four feet above the roof line and a foot beyond the edge of the building when the roofer was injured, the code when originally installed over 30 years earlier required 8 feet clearance and current code required 12.5 feet. Moreover, I demonstrated that despite the power company's claim that they were unaware of the too-close sag, they *must* have been aware because they had surely retensioned the cables at some point in that 30-year period. The evidence was that the currently deeply-sagged cables were not long enough to have been attached to the original straight poles. I also made a model to demonstrate all of this using floppy bead chain for the cables and pivoting poles to show the inconsistency between the cable's length and the originally-straight poles. As a result of my findings—reported in a written document and supplemented via video demonstration of the model—the power company agreed to GTE's paying a smaller fraction of the settlement *and to* revising the tenancy agreement to eliminate that "1% liability equals 50% cost sharing" provision.

In my last calendar year at GTE, my division was disbanded as were all other hardware divisions. Naturally, I was concerned about my tenure, but managed to pick up a position in the **Network Security group**. Being a software-heavy group, it was not a good fit for me, but I kept busy helping with a demonstration lab, being a local liaison to the specifications-writing team in Texas, and modeling some router networks.

As soon as the 90-day notice period arrived for giving notice of early retirement, I did so.

It was a great nineteen years, but it was now over.

OSRAM SYLVANIA
November 17, 1997 to January 31, 2005 – Ages 57 to 65

As my early retirement from GTE approached, I planned to become a consulting engineer. Hence the creation of ABell Engineering that consisted of some business cards, stationery, and a DBA from the town. However, those plans quickly went into abeyance when a former colleague at GTE who was now a manager at Osram Sylvania Inc. (OSI) heard of my news through a very rapid grapevine and called me about an hour after I tendered my retirement. The opening sounded interesting, so I somewhat reluctantly went to their office for interviews. Surprisingly, every time I rounded a corner and saw a familiar face from six or more years earlier, I remembered the person's name. That, alone, "must be a message," but I also liked the prospect of the work, the environment, their acceptance of my working a four-day week, and the prospect of sharing the 33-mile commute with a neighbor. I took the job. In view of how the stock market tanked over the next decade, taking that job was certainly fortunate.

Over the next three years I worked on several challenging support projects. One was the design of thermally-efficient ballast castings—two or three of which were machined in prototype form and tested to be effective. That same ballast group occasionally asked me to design miscellaneous product hardware or laboratory equipment. For a while, I was overseeing two interns from Germany and managing their analysis of fixtures designed for the recently-introduced Icetron lamp. The materials group called on me to design a small vacuum chamber that included a full-size viewing window and articulating measuring and effecting mechanisms inside for processing samples under vacuum. Also for the materials group, I designed a centrifuge for testing lamp filaments under stress, and did extensive design work for the bulgy arc tube project. One summer, the quality assurance group asked my advice on saving a 10-foot diameter integrating sphere, which I did in a few weeks and saved the replacement cost of $125,000. A brief project to design a pressure seal between two tiny pieces of steel resulted in a subordinate position on the only patent from Osram Sylvania with my name (see US Pat. No. 7,052,649). Another summer, an intern from WPI worked under my supervision to design and build a *Data Tally* for remote monitoring of prototypes deployed in the field. And for a plasma-study group, I designed a complex floating bell jar system for rotating an arc lamp in or out of synchronization with a thermal video camera for thermally mapping the effect of gravity on the arc. Some of these projects are described in more detail below.

Icetron Lamp: The Icetron microwave induction lamp began at GTE Labs and had recently become a product at OSI. It was a marvel: about the size of a flattened loaf of bread, cool white light, super efficient at about 76 lumens per watt, made in 100 and 150 W sizes, but it had a solid-state ballast that simultaneously needed to be kept cool *and* mounted within a foot of the lamp. Most of the initial customers made fixtures consistent with the size of the lamp instead of the size of its heat output. The fixtures frequently overheated and caused the ballast to cycle on and off. Many early installations were disasters. In fact, the problem was so acute that marketing eventually required customers to have OSI test their fixture before agreeing to sell lamps to them.

Bill and I were handed this responsibility of developing a qualification program that was supported by a succession of three or four university interns from Germany. Many hours were spent testing these fixtures, and most failed the first time around. We made suggestions for improvement and sometimes introduced the improvement ourselves and retested. Bill and I made several field trips to customer's shops to give advice. And a huge Sunoco sign was delivered to our parking lot for designing, installing, and testing a sample. Eventually, sales of the Icetron became very successful.

Evacuated Centrifuge: The centrifuge was designed for a materials engineer whom I had known at GTE. It was to be used for testing the sag resistance of lamp filaments burning at full brightness while under the stress of increased "gravity." The vacuum was needed to purge the air and simulate the atmosphere inside a light bulb while the spinning increased the apparent gravity. This was an enjoyable project. He was developing new alloys of tungsten filaments and needed to test multiples, so I fit in about ten filament stations. The samples needed to be changed-out quickly and easily in a factory setting, *i.e.* low skill, but without distorting them before testing. As filaments are highly flexible, avoiding stretching or compressing them was accomplished with an indexed cradle to hold it under neutral stress as it was set in place. Ease and speed were accomplished with a simple pincher at each terminal that was balanced to tighten slightly with increasing centrifugal force.

Power was substantial and needed to enter through robust rotating slip rings, ferrofluidic seal, and vacuum-tight feedthroughs. The considerable thermal load needed to be conducted outward for dissipation. I coordinated the design with a vendor who specialized in vacuum systems so that my design included a stainless steel bell jar-like cover and a baseplate with all the testing hardware, drive motor, and feedthroughs installed. The vendor mounted the baseplate on a stand with vacuum pump, gages, and an air-spring lift for the cover. It worked so well that a second unit was ordered for a different factory.

Bulgy Arc Tube: When I joined OSI, the materials group was designing an arc tube made of ceramic. It was lovingly called *bulgy* because it bulged in the middle. The bulge was a ball about half an inch in diameter, and the electrodes entered through long, narrow tubes at opposite poles. The material was mostly alumina (aluminum oxide) that when fired was 97% transparent although it was not clear like glass. An outside vendor had made some prototype tooling to make similar shapes—my job was to design molds for in-house prototyping of the bulgy parts in moderate quantity. I worked closely with Scott, a likable and practical PhD throughout this multitask, multiyear project.

One part of the design was the injection system, which required some straightforward design work. But the challenging and interesting part of the design was the hand-held mold. This was made as a cylinder about the size of a Campbell's soup can, but in five stacked pieces that could be disassembled in a prescribed sequence to eject and remove the delicate, still-green part. The molded part was one half of the finished arc tube, and two of them would subsequently be sealed together at their equator to make a whole tube. As molded, the parts were 134% of final size—they shrunk greatly when fired. I also designed fixtures for opening the mold in a uniform manner without the risk of squeezing or otherwise damaging the still-delicate part. Holders for assembling two halves had to be designed, and I also devised an improved shape for the mating edges that obviated entrained air pockets.

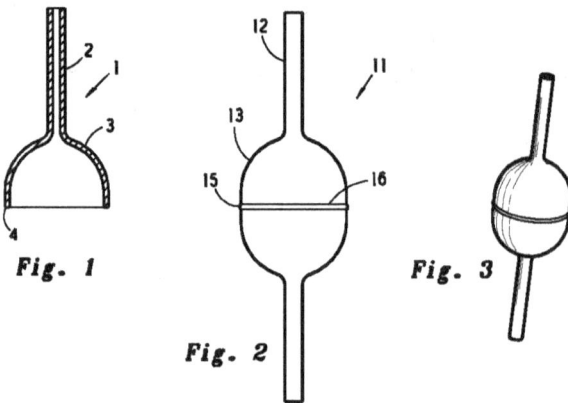

A drawing of the bulgy arc tube in three stages of completion including the shrunken third stage after firing. Drawing made by the author for the team to use with a design patent. Only the perspective image at right made it into US Design Pat. No. D484,255 "Ceramic Arc Tube."

Eventually, I designed 10 or more molds for different sized tubes ranging from 35-watt, the size of a green pea, to 400-watt, the size of a cherry tomato. Each of these molds had concentric rims for alignment, vents to let the air escape, ejector features, a pin to form the ID of the electrode section, sprue cutoff, and a high polish of the interior. For each bulgy, the CAD system helped me calculate the final interior volume to help establish an optimum fill when making the finished product. This project stretched out over several years and in the final stages resulted in consulting payments because I was then getting my regular salary from a German division, not the project-chargeable local office. Later, I had the satisfaction of seeing an automated version of the same production process that used robots to handle the molds, which was operating in the New Hampshire factory.

Salvaging a Sphere: A huge integrating sphere at the Manchester, N.H. factory was used for routine testing of 1,500 watt metal halide lamps—a large version of the 175 W lamp tested in GTE Labs' first GAS payload aboard the Space Shuttle in 1984. The sphere was ten feet in diameter, constructed of steel in two halves split vertically, and hinged to open like a clam's shell on rollers for inserting a new lamp to test. The interior was painted white so that

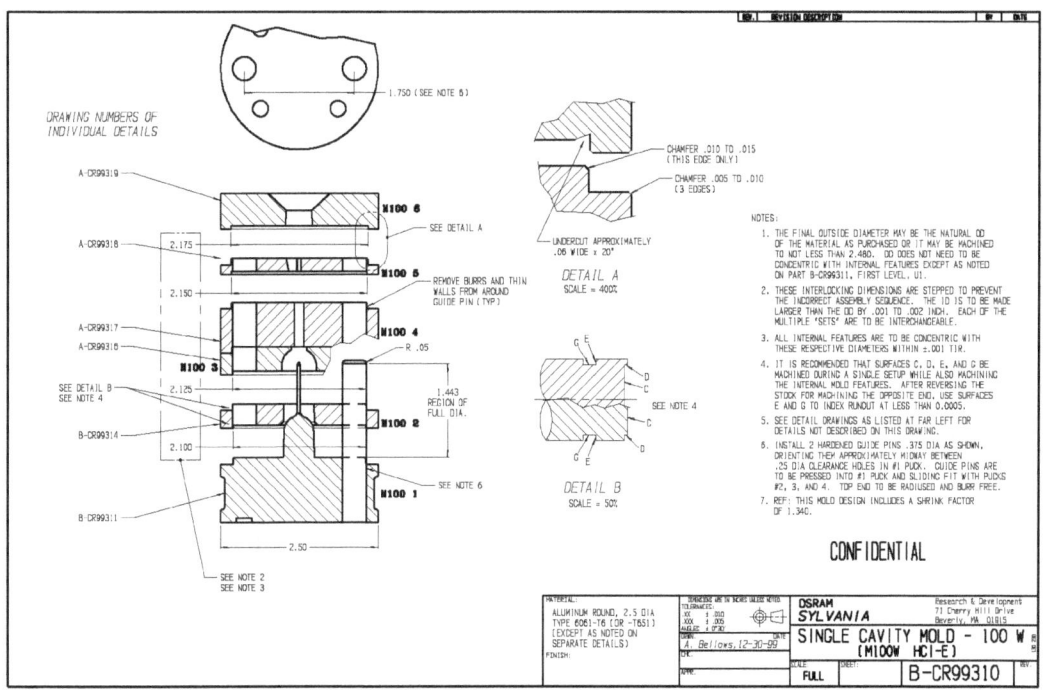

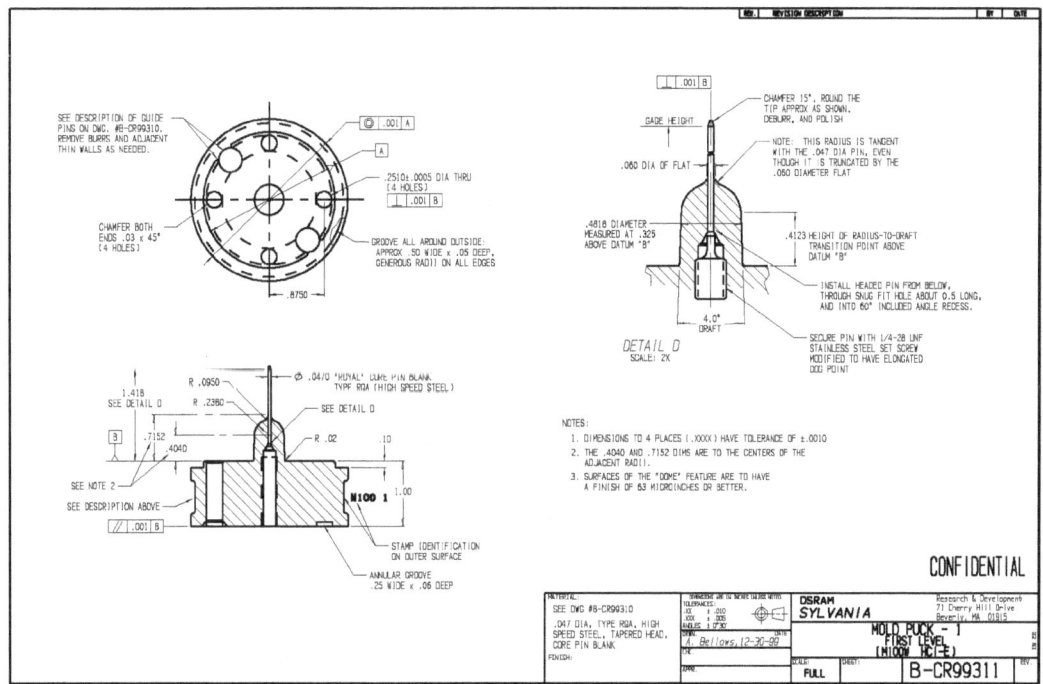

Two of seven sheets of a detail drawing of the hand-held mold for the bulgy arc tube. Drawings made by me using Cadkey, my favorite CAD system. (Reduced here to 33%.)

the light from the lamp in all directions became completely mixed or *integrated* before being measured by a photocell. The moving hemisphere had squashed out of shape enough that the sphere no longer closed properly, resulting in bogus measurements. A new one had been quoted at $125,000. The quality manager asked me if the sphere was salvable.

After staring for hours at this beast constructed of curved angle irons and heavy hemispherical sheet steel shells, I surmised that the primary culprit was that the 2-inch diameter hinge pin had been cut into two sections years earlier, presumably to resolve some other problem. I made no attempt to replace the pin, but added an adjustable brace to each hemisphere at the hinge knuckles and designed new casters with screw thread height adjustment.

The parts were made to my design by the factory's shop, and during the July shutdown, I spent most of the week with a maintenance man installing the new casters and braces. After adjusting the wheels and the two braces with their jack-screws, the hemisphere eventually came into shape for a perfect fit. I told the quality manager that surely he owed me half of the saved replacement cost. He offered to pay me in light bulbs.

Data Tally: Management asked me, "Do you have a project for a student engineer to work on for ten weeks?" I proposed a temperature, power, and light output logger with telephone access that could be used for field testing of prototype products, such as the Icetron. The manager thought it too ambitious but ultimately agreed. I acquired most of the anticipated parts based on the Tattletale TFX-11, the same Cermetek modem used in the Back Watch at GTE, and some Analog Devices thermocouple amplifiers.

Summer arrived and Leo showed up to begin. The name "Data Tally" seemed snappy to me, and he agreed. I taught him the ropes of using the Tattletale development system, offered to create the printed circuit board artwork if he laid out the components, and helped him find and buy the other components. He built and tested a clip-leaded breadboard, sketched out the part arrangement, drew the schematic, wrote and debugged the BASIC program, and I generated the artwork and paid extra for a three-day turnaround of PC boards. Within the ten weeks he finished the assembly, wrote the software, and demonstrated a device that collected temperatures from five ports, supply voltage and current at the product, and light output at selectable rates, store those data along with a time stamp, and periodically deliver results by telephone. It was a happily-surprising success for us all.

• • •

The LED Business – In the summer of 2000, a fellow Member of Technical Staff and I were invited to transfer to Osram Opto Semiconductors (OOS) based in Regensburg, Germany, although our offices would remain in the same place. Our new responsibilities would be to design and apply lighting products for the US market using light emitting diodes: LEDs. I was having fun at OSI, so was somewhat reluctant, but thought, "Maybe it's time to try something different." The other person was Joe, and we would be reporting to a manager in Regensburg, where we visited twice. We would have a local manager: Chips, an organic LED chemist who was now in marketing.

LED PROJECTS – The OOS group in Regensburg had designed a collection of LED modules: a *LinearLight*, a *BackLight*, an *EffectLight*, and a *MarkerLight*. Each of these comprised multiple LEDs in red, yellow, green, blue, or white mounted to a circuit board and designed for operation at either 10 or 24 V, depending on module. The circuit included the necessary current control device needed by LEDs. Ostensibly, we were to design US modules, but our initial job was to help Chips with his marketing of these modules, and to that end, our tiny lab was soon stocked with multiple cartons of every product in the collection. Joe and I worked as a team for the next four years.

My first assignment was to outfit a small sign supplied by a customer with LEDs as an enticement sample. One common version of signage is installed as a sequence of individual letters, called *channel letters* that are usually backlighted with fluorescent lamps or neon tubing—think of WALMART with those huge glowing letters nailed to the wall across the front of the store. This customer had sent us a single channel letter about 18 inches high and constructed as a sheet aluminum box with a red Plexiglas face. The BackLight was specifically designed for this type of application.

This seemingly simple project had a number of challenges—how to fasten the 1-inch square circuit boards of the BackLight to the aluminum back wall, how to be sure the circuit traces and delicate flexible leads would not short to the aluminum, how to connect successive strings of BackLights together, and how to power them with 10 volts DC. Solutions were clumsy and time-consuming. I cut out a piece of corrugated cardboard to fit inside for an intermediate back panel, attached the BackLights to the cardboard with stubby screws, carefully soldered the tiny leads together, then insulated them with shrink tube. The power supply was the trickiest because we soon learned there was no such thing as a 10-volt supply sold anywhere in the US. Because this was a one-time demonstrator, I outfitted a 12 VDC Radio Shack wall adapter with three diodes in series to drop the voltage closer to 10 V, soldered the wires, and shipped it off. All this took hours and looked distinctly amateurish. Apparently the customer wasn't impressed with this LED-lighted sign and was never heard from again.

This three-day exercise in futility was prophetic, but I didn't recognize it at the time. The remainder of my four years with OOS was spent fighting the same issues over and over again. Three of the four module type products were designed to be connected by soldering your own wires to tiny $1/16$-inch square pads without holes for anchoring the wire—a process that required three hands and resulted in a weak joint. The BackLight had wire leads, but they were so short they were hard to hold while stripping, and the wire diameter was so small that no US-made wire nut could secure them. Two modules had tiny holes for #4 mounting screws, but the screw head would short to the adjacent circuitry, and an added nylon washer would crush down on a nearby component. Two other modules had no mounting features whatever. And, as already stated, there were no power sources available for the 10-volt products.

In the early 2000s, LEDs were not very efficient—only the red LED approached the efficacy of an incandescent bulb, which is only 15%. The others were less. Our customers generally thought that a solid-state light source must be very efficient, bright, and cool. They were none of these. Oftentimes, we found that customers only wanted white LEDs. Actually, LEDs don't come in white. They use blue LEDs, the least efficient, covered with a phosphor to generate a broad spectrum of red and yellow with some blue peeking through, thereby simulating a somewhat white appearance, but at an exceptionally low efficiency.

The EffectLight and four sizes of MarkerLights were marvels of optics, but had virtually no applications among our customers. The LinearLight was a $3/8$ inch wide rigid strip about 17 inches long with 32 LEDs, and the BackLight was a wire-connected 8-unit string of 1-inch square circuit boards each with 4 LEDs. The BackLight drew interest from the signage community for channel letters, and the

LinearLight drew interest in the decorative lighting community. However, both of these potential customer groups generally lost interest as they learned first hand how dim, expensive, and difficult to apply they were.

So, what did I do during my four-plus years in this business?

For one, Joe and I both spent a lot of time **on the phone** and shipping out samples. Some questions were so frequent that eventually, I composed a set of tip sheets, each on a single topic that could be sent to a customer in need. "How do I solder these LinearLights?" Send *Tip Sheet 4*. "Do you have suggestions for powering and mounting BackLights?" Send *Tip Sheets 3 and 8*. Each tip was a single page and included illustrations and schematics drawn in my CAD system and integrated with the descriptive text. They were routinely packed with samples, mailed, or emailed to all who asked.

We also assembled many **sample demonstrations**. The first was that small channel letter described above. Joe undertook outfitting a 4-foot high channel letter "A" for Walmart with white LEDs. He spent weeks making an insulated subpanel and installing about 75 modules crammed in along with hundreds of delicate solder joints and an unbelievable scramble of wires. After all that effort and even with 2,400 LEDs, the letter was much dimmer than it had been when delivered with only 5 fluorescent tubes. After Chips saw the result and estimated that the total parts cost for a single WALMART sign would be about $20,000, he didn't even invite the customer to see it. I spent many months working with a midwest sign maker trying to develop a satisfactory canopy border for Shell gas stations that included both red and yellow LEDs. As designed, our LinearLight could not quite satisfy the specs. (That sign was eventually introduced, but not with our LEDs.)

I spent over a thousand hours trying to **resolve the power supply** issue and getting Underwriters Labs (UL) approval of our products. As noted above, 10 volt power supplies just did not exist. I got quotes for making custom units for us, but that plan went nowhere. UL considers 10 and 24 volts to be dangerous enough that power supplies in that range must be rated as *Class 2* (*e.g.* power adapters, chargers) and limited to about 6 amps for a 10 volt unit. Six amperes would not light many of our LED modules. For example, that 4-foot high sample for Walmart needed five separate Class 2 power supplies for the one letter. After we cast about for more than a year, I discovered that Philips made a neat 12 V, Class 2-rated power supply for their LED modules. Although a competitor, they agreed to make us a private-branded 10 V version, and sold us both 10 and 24 V look-alike units. An outgrowth of learning about UL Class 2 issues was the also-unexpected discovery that the modules themselves needed UL approval. I spent nearly a year working all of those products through the approval process. I also acted as liaison between a British company and UL to get their green LED system UL-approved for BP gas stations. For that effort, I was rewarded with a bottle of Glenmorangie single malt Scotch whisky!

Some of my time was spent **fielding impromptu questions** about unusual applications, and I designed many custom electrical wiring configurations to optimize the voltage maintenance of these relatively high-amperage applications. For example, I devised a highly effective (but confusing to most customers) wiring configuration that ran the +10 V bus from east to west and the neutral bus from west to east so that voltage drop along the wires would cancel. For many applications with this arrangement a 12 V power supply was ideal because the voltage drop in the wires could be tailored to be 2 volts over their length, supplying 10 V to every module along the bus even with small gage wires.

This configuration was used for many customers. To optimize them, I modeled each application using a spreadsheet to calculate current flow and voltage drop given wire length, gage, and resistivity. After several such custom models, I generated a more generic model that could be implemented for most configurations with a few simple changes in boundary conditions. This was very useful and was later used for the Jefferson installation described below.

Another undertaking of mine was **designing accessory hardware**, primarily for sales purposes. One example was a custom sems screw with an integral and undersized nylon washer that would have resolved those many requests for mounting screws. It was self-tapping and would therefore be acceptable in metric countries as well. The minimum order was 25,000 screws and the $1,100 was never approved because supplying special screws might "call attention to a problem."

Another example was an easily portable display case for demonstrating all the LED modules. I used an ordinary Samsonite brief case to carry a two-sided custom panel embedded with two of our power supplies, outfitted with a dozen power sockets, and covered with the fuzzy-side of Velcro. Each of the included modules had patches of the hook-side of Velcro. Thus a sales rep could arrange the modules to meet a particular customer's interest, and the modules could be removed for closer examination or to place into a display using the included extension cord. I devised the power connections to use short sections of telephone coil cords and modular connectors with 4 contacts—one pair for 10 V, the other for 24 V so there was no need to select sockets. Modules could populate both sides of the panel that could also be removed completely from the briefcase to stand up on a table or be temporarily included in a wall display. My technician built 20 units. Finally, some success—too bad it wasn't a sellable product.

For two displays, I programmed Tattletale-based microcomputers to cycle the lights in various modes and to demonstrate dimming. I also repackaged the control circuit from a toy to change the colors of a newly-introduced color-adjustable LED LinearLight.

The only time that Joe and I were asked to **develop a module** for the US market—our originally-promised responsibility—was in my fourth and last year in the LED group. A task force of four including Joe and me visited and interviewed dozens of customers around the country about how they would like to deploy a new, high-brightness white LED. Every customer had a different opinion. After the interviews, our team attempted to boil down our findings to a meaningful product description. Our final spec didn't match any of the customer's requests, but it was a moderately good compromise.

The design ended up as a flexible substrate about 18 inches long, $3/8$ inch wide with 1-inch bulges every 3 inches for the high-power LED mount. The ultimate intent was to manufacture it as a continuous strip that could be cut with scissors at any of the 18-inch increments. It included a number of features to resolve many of the problems we had encountered over the previous years. Since the LED was called *Dragon* by OOS, and the long flexible circuit strip reminded us of a whip, we dubbed it *DragonWhip*! I laid out its details, had a circuit designer convert it to PCB software, and ordered the substrates. I was able to assemble a few samples before my retirement and left detailed assembly and testing instructions.

One LED project includes a poignant connection. In 2001, Osram Sylvania made a gift to the Nation—a major overhaul of the lighting in and around the **Jefferson Memorial**. One component of that was the installation of hundreds of LinearLights in the rotunda 40 feet above the floor on a narrow 3-inch ledge beneath a quotation from Jefferson. The contractor called on me many times for help with supplying power to this large system in tight quarters. He applied my "east-west" configuration with success. But with only a few days to go before the grand opening, the installers called to report that the adjusted brightness of the modules was spotty.

I flew down to Washington with tools and meters in my briefcase to see and hopefully debug the problem. I boarded a lift and went up into the rotunda close to a randomly selected group of modules mounted in fixtures. To my surprise, a mixture of *Version 1* and *Version 2* modules had been included in the installation. This would not have mattered in most installations, but this specific application required dimming, and we had *expressly* ordered modules of only one version to be delivered so that the dimming properties would be uniform throughout. Fortunately, Joe was able to find and deliver the needed modules overnight, and the installer worked all weekend to change out the odd modules.

My visit was not just any Thursday; it was Thursday, September 6, 2001. While there, I also climbed to the top of the Memorial and could see, among other things, the Pentagon across the river. The following morning I returned to Boston while sitting on the port side of the plane. The air was pristine, and there were beautiful views of the Pentagon and the two World Trade Center towers in Manhattan. All three of those would be attacked four days later on **September 11** and two completely destroyed. Needless to say, the grand opening of the newly-lit Jefferson Memorial was postponed to a more modest occasion several months later.

January 24, 2005 was my last day in the office. There was a generous going-away luncheon organized for me at which some unexpected old friends showed up. I gave a little thank-you talk and summary of how much I had loved my engineering life, the people I had worked with, and how fulfilling it had all been.

• • •

In the dozen years since I left LEDs behind, there has been vast improvement in LED efficacy that makes them glow brighter and run cooler. Clever integrated designs of LED-based products have rendered them far more effective than trying, as we did, to adapt previously-designed modules to new applications.

HOMES, RENTALS, and REMODELING

As stated earlier, my way of life has always been to bring my technical, and therefore professional, habits home with me. This is not to say that I brought "homework" with me—in fact, I rarely worked at my day-job while at home, but I did continue my practice of isolating problems, resolving them, and creating solutions. Moreover, each of the houses owned, occupied or not, was so rife with challenges, problems, and projects, that I feel compelled to describe some of my experiences with them, technical and otherwise.

The Cambridge Apartment House – As reported in the *Outside Polaroid and ECA* chapter, I made my first real estate purchase in 1968 when 28 years old. I will not repeat that story and my experiences with that watershed event other than to remind the reader that I put more than 2,000 hours into remodeling, repairing, and generally improving that ancient structure, part of which was my home. It should be noted that I completely rebuilt one kitchen and partially rebuilt two others for an equivalency of two kitchens. Also, this house was an outstanding investment. Depending on how one computes ROI in the complex scenario of borrowed money and ongoing income, it returned some 12 to 16% annually. In addition, I lived for *free* in one of its units for six years.

The Belmont Home – As our family began to grow, we decided to establish a home but keep the Cambridge apartment house as a rental. In 1973, we found an old ark of a house in Belmont, Mass. with a lot of eccentric features that appealed, including a relatively large lot. We didn't move in for another year because of an existing lease.

This 1841 Greek revival faced north, providing a sun-drenched back yard. Reportedly, it had been enlarged soon after the Civil War and appeared to include some remodeling in the twentieth century. The 16-room house was over 80 feet front to back. Most of the basement floor was dirt and some of the floor beams above were of tree trunk size. Several rooms on the first floor still had massive doors and trim that had been hand-painted to look like oak. I found that detail a bit too garish, but was too protective of history to paint over it. Generally, rooms were large, especially the 16 by 32 foot living room.

At the rear was a kitchen complex including a large kitchen, a butler's pantry leading to the sizable dining room, four closets, stairways both up and down, a large storeroom, and a back hall with a rear entrance, closet, and small bath with only a toilet of the overhead tank-and-chain variety (the Thomas Crapper style). The kitchen floor was of horrible old linoleum, and the tenants had recently painted the kitchen white with orange trim.

The second floor was also rambling with large bedrooms at the front, medium sized rooms in the middle, and smaller rooms at the rear, undoubtedly used as servants' rooms at one time. The third, attic, floor had two finished rooms at the front and a huge unfinished attic space crammed with old furniture and junk.

Only two basement rooms had cement floors, one beneath the kitchen, and one beneath the rear-most storeroom and back hallway. I used the latter for my shop as it had a higher ceiling, more room, a window, and direct access to the yard through a single step-up door.

The yard had been largely neglected for many years; about half of it draped with dense grape vines and strangled trees supporting them. The land dropped by a full story toward the rear of the house permitting the walk-out basement door. Immediately behind the house was a huge, beautiful maple tree. In the winter I discovered a dry-laid stone wall under that jungle of vines that dropped the land another two feet to what had once been a tennis court.

• • •

So, what did we do with this big old house? Work, work, work. One of my first projects was replacing the door locks with interchangeable cores with master keys for us. Therefore the cores could be switched around easily for tenants or workers as they came and went.

During the first four years, the upstairs front area was rented out: two big rooms, a small room, and bath. We typically used the side entrance and hardly ever saw the tenant. After the first year, I installed a kitchen in the small room making the area into a decent living space. Soon after starting work at Foster-Miller, I began to rent the two back second floor rooms to a fellow employee who sometimes pitched in to help me with projects.

During the first spring and before moving in, I cleared a section of the side yard to create a garden. The topsoil must have been at least two feet thick and was always moist, presumably from water seeping down from Belmont Hill across the street. The garden was a great success; everything flourished, including the weeds.

This house and yard required unending work in the five years I was there. In time, it became truly tiresome. For two summers, I hired a college student to work on both inside and outside projects. At night I would write up the project for the next day along with diagrams and leave it for him to follow. The next evening, I would evaluate his progress and write up the next step or another assignment. Many projects got accomplished this way with little effort on my part other than planning and purchasing supplies.

The side entrance had a crudely-added and ugly enclosure that obscured some handsome corbels for the roof above. I tore the walls out, reset the brick walkway, and replaced the adjacent weeds with greenery.

The monumental project of redecorating that cavern of a living room: stripping the wallpaper, washing calcimine from the ceiling, and sanding, cleaning, and painting the miles of baseboard, chair rail, crown, trim, columns, and cabinetry took many weeks. It looked so fresh and bright but that huge space was never adequately furnished or comfortable.

• • •

Only one bedroom had its own bath, thus it was adopted as master bedroom although it had no closet. I improved its bathroom by adding and tiling a shower to the tub, and switching to a high awning-style window that obviated the need for shades and gave space for a towel bar below. The toilet and the pedestal sink were funky but in good condition, so I left both of those in service.

SURPRISE CLOSET STUMBLE – Adding a closet turned out to be a much bigger undertaking. There was an obsolete hallway that separated the master bedroom from a second bedroom. The plan was to convert it into three closets: one for each bedroom and a linen closet opening to the rear hallway. I started this project on the hottest day of the decade—103°. The first phase was to remove all the plaster and lath for new doorways, so I brought in all the needed tools and a dust mask, and taped myself inside with plastic barriers to keep the dust under control. Several hours later I emerged white all over, retaped the plastic, and took a long cool shower.

But I discovered a big problem. The section of wall planned for the new doorway contained a huge truss comprising two diagonal 4x6 timbers and a $3/4$ inch diameter steel rod, all of which supported the 8-foot wide sliding door to the dining room below. I also discovered a previously-unnoticed section of baseboard that was removable to give access to an adjustment nut on that rod. I wondered: had anyone else ever noticed that section of removable baseboard over the previous 134 years?

I relocated the closet door a few inches to avoid the steel rod, installed a steel channel-beam at ceiling level within the future closets to bridge the width of the sliding door, attached the steel to the upper end of the truss with bolts, and removed the one diagonal timber that was blocking the new doorway. I opened, framed, and cased the new closet doors, installed two walls across the hallway, and ended up with two new walk-in closets and a linen closet. I also adjusted that nut to square-up the sagging doors below.

• • •

When we first moved into that house, the kitchen had no convenient workspace or storage except for all those closets and the storeroom. I made a short-term improvement by temporarily placing some of the newly purchased cabinets where needed and outfitting them with a loosely-placed vinyl-covered plywood top.

KITCHEN – A couple of years later, the serious remodeling of the kitchen began. Like the kitchen in Cambridge, this one also had doors in every corner, but the revised layout included tearing out some walls to create an eating area. The teardown removed two of the doors. The final kitchen included a beautiful refinished fir floor; a new oversized casement window over the sink; a bay window in the eating area; refinished natural fir trim; the original ivory and green enameled 1920s-vintage gas stove; one base cabinet adjacent the freestanding stove; a recessed safe hidden behind a movable cabinet; a row of cabinets that included a stainless sink and dishwasher; and generous wall cabinets above. I fabricated the Formica countertop in white.

In the eating area, I built a bay window to be five feet wide with a built-in planter. That project was so well organized that in a single weekend, I used the chainsaw to cut the opening in the outside wall, carefully cut the clapboards back to align with the future siding, installed my prefabricated, sag-resistant end panels onto the wall using lag screws and carriage bolts, stick-built the outer wall and the bottom closure, stick-built the rafters and installed plywood across the top for a sloping roof, installed the two crank-out Anderson casement windows at each end, built a temporary five-foot wide ladder, and with a neighbor's help, lifted the heavy thermopane picture window up that ladder and into place. But the pride of all that accomplishment was negated the next weekend when I was only able to lay the few roofing shingles and somehow got them wrong. The planter was lined with laid-up epoxy and glass mat and outfitted with a drain to drip out the bottom.

This house got my third and fourth kitchens.

• • •

In the summer of 1976, my Foster-Miller tenant helped me build a 12 by 14-foot deck directly behind the house. The rear door had previously led to a rickety stairway down to the ground a full story below. I removed that stairway, repaired the siding, and constructed a well-elevated deck with a two-section stairway to the ground. The railing along the far edge was built with a slight outward cant and a deck-length bench seat. It was a great deck that got used a lot, and because of its height above the ground, insects were rarely a problem.

I lived in that house for five years, the first year married to my first wife, three years as a single Dad, and the last year with my second wife. Most of the projects described above, in addition to a lot of others, were completed in the middle three years. My daughter typically came to stay for one night every week and every other weekend. She made good company when I worked. Sometimes she followed me around, sometimes she had her own projects, and sometimes we went on excursions elsewhere.

The Waltham Apartment House – In 1976 I felt it time to invest in another apartment house. The Cambridge building was largely taking care of itself, so I assumed that another should work for me just as well and perhaps better if it were without rent control. The new job at Foster-Miller was not exciting enough to capture my full attention, and I needed another diversion. I looked in Waltham, which was close to both home and work and had a lively rental market.

The building I found had been built a century earlier as the Kilby Hotel; had been converted for 12 apartments at some time; and had recently been remodeled with new siding, new entries and doors, and new baths and kitchens. It had a coin laundry in the basement, and generally seemed to be in good condition and a good buy. It took a little doing to pull the financing together without assuming the 18% second mortgage that the seller was offering.

Unfortunately, it turned out to be a nightmare. Between the time of my offer and the closing, four of the twelve apartments went vacant, and I had a struggle finding new tenants while throwing in money just to keep the payments current. Several tenants were troublesome with constant complaints or tardy payments. The basement regularly flooded an inch or two even during mild rainstorms, thus the coin laundry was rarely used. There was inadequate hot water, and several of the 12 boilers needed repairs. About half the tenants had been issued a master key, so there was grumbling about security from those in-the-know. Depressed over the place and desperate for some way out, I put it on the market for nearly a year but got no takers. But the situation also provided some interesting characters and entertainment along the way.

While still considering the purchase, I visited the place one evening with my little daughter in tow. Inside the hallway, she remarked, "I don't like this place, it stinks" referring to the smell of cat urine. You cannot believe how many times over the next two agonizing years I thought of that prophetic statement from the mouth of a four-year-old.

• • •

The person who renovated the building a few years before my purchase was a plumber. The plumbing should be in top shape, right? It wasn't.

PLUMBING – The most expensive surprise was that although each gas meter was correctly billed to the tenant of the respective heating boiler, the pipes to the kitchen ranges were largely scrambled and in some cases doubled-up. This was discovered when one of the tenants stopped paying his gas bill, had his meter locked off, but continued to use his oven to heat the apartment. The tenant below guessed what was happening when his bill spiked. The plumbing crew contracted to unscramble the pipes spent nearly a week adding and reconnecting pipes at great expense.

Another pipe story: A tenant reported a leak through the ceiling from the bathroom above. I checked it out and after several ineffective repairs of the tile work, eventually traced the problem to a leak in the pipe between the control valve and the shower head—hence only occasional leaking. The upstairs tenants gave me permission to inspect their plumbing through the easily-patchable bedroom wall, which I did during my lunch break one day. I poked a hole near the valve: I could see drops of water from above. Poked a hole near the showerhead: no drops seen. I poked more holes until homing in on a leaky coupling in the pipe. A plumber was needed, so I went back to work leaving five or six holes in their wall. That evening, the wife of the upstairs couple called to ask, "Did you do anything in our apartment today?" I described my leak search. She was so relieved. She had thought those holes were a *"sign."* Apparently, she was very superstitious and thought that some kind of spirit had entered her bedroom and made these marks as some kind of *message!*

I didn't need such aggravation—I had leaks to fix. The plumber cut around and gave me the bad joint to dissect. I quickly patched and painted over the holes for these troublesome tenants. I also carefully disassembled the joint and discovered that the solder job had been terrible with very little solder actually joining the two sections of pipe. Minor corrosion was all it needed to begin leaking. Before I sold that house, three more similarly bad joints gave way resulting in costly repairs. So much for the expertise of a professional.

Though not the plumber's fault, there was a surprising problem with a toilet. A tenant called to report a clog they could not clear with the plunger. I snaked it open. And then again a week later. On the third call, I showed up with tools and supplies for pulling and resetting the toilet. Wow, what a surprise. The three-inch diameter flange that extends down into the drain pipe horn when set in place had a tiny defect in the glazed clay in the form of an abrupt recess about $3/16$ inch in diameter and $3/8$ deep. A Q-tip had found its way into that little hole, caught there, and formed a rod across the pipe on which paper could build up and eventually block passage entirely. The plunger and snake had only cleared the blockage, but had not plucked out the Q-tip. I did that by hand. No more trouble.

STUMPERS – That flooding basement was a real problem for me. I often needed to go fix some trouble in the basement. If it had been raining any time in the past week, it meant getting my shoes soaked. I went there many times during the rain in hopes of finding the source. It was either still dry or already flooded. I had several theories, repaired each, but to no avail. Finally, on one of those visits I was there when it started. Quite suddenly, a stream of water emerged through the loose-stone foundation at floor level and began to flow across the floor at a prodigious pace. It reminded me of the tidal bore in Nova Scotia—looked minimal to start with, but suddenly there was all this water everywhere! I grabbed a rag and a broom handle to try stuffing up the leak. The leak just moved over by a foot. The solution: I broke through the concrete floor, installed a sump barrel, formed a shallow dam of mortar that looped outside the barrel and along a ten-foot stretch of the leaky wall, and installed a sump pump. Dry basement ever after. Soon thereafter, I replaced the clunky washing machine, notified everyone of the new and dry laundry, and could not believe the revenue. No wonder the Mafia operates so many laundries.

And the insufficient hot water? No surprise: it was only a 75-gallon heater—residential size, and this one served about 20 people. A larger one meant an industrial type, which meant a big expense. So, I bought another 75-gallon heater, plumbed them in series, turned up the temperature, added a "tempering valve" to regulate the outflow temperature to a reasonable level, and never got another complaint.

Whenever I showed or described a vacant unit, many potential tenants rejected it because the kitchen was just a row of cabinets and appliances at one end of the living room. They were perfectly nice, but the arrangement was bad. One day I saw a cartoon in my boss's office showing a shack in a swamp along with its real estate ad extolling the virtues of a water view. Hmm... Thereafter, I described the unit as having a Mary Tyler Moore kitchen. Rentals greatly improved.

So how did this story end? As the two-year ownership was approaching, the physical problems were settling down; vacancies were filled; some of the new rentals were at higher rates; revenue was increasing; and the tenant problems, mostly collecting the rent, had been taken over by my fiancée who was better-equipped to deal with disagreeable personalities than I. In fact, it was going so smoothly that I took it off the market. Not long afterward, a stranger called and told me he had seen my building, thought it would be just right for a family investment, got my name from the tax office, and asked if I would sell it. I didn't want to seem anxious, so told him I would think about it and took his phone number. A few weeks later it was sold. I had owned it almost exactly two years, and the closing just a week before our wedding made a nice wedding gift.

As an investment, this one was poor. I came out "whole" but it did not begin to rival the return eventually had by the Cambridge apartment house.

Our Home in Wayland for 29 Years – In 1979, the year after we got married, we moved to the suburbs. We had looked in several towns including Wayland, Massachusetts, and this house settled out as our choice.

Historically, it was a Dutch Colonial style house built about 1924 on a tiny lot. Our predecessors had bought additional land to make the lot slightly over an acre and constructed a sizable addition in 1964. The addition extended to the left of the original house and included an in-law apartment on the first floor and a huge master bedroom on the second. With the addition, the house had a total of six bedrooms.

There was a separate two-car garage with paved driveway.

We had no need for the in-law apartment, so we used its living room as a family room, one bedroom as my office, and the kitchen as a mudroom.

Early projects in that house included redecorating the living room and dining room, removing all the smoke-scented carpeting from the in-law area, and painting or papering over its institutional-green walls and woodwork, all of which my wife did.

Two major projects were undertaken the second summer we were there. One was to build a brick patio in the back yard behind the kitchen. The other was to remove the ancient steam boiler from the basement, remove all the pipes and ugly radiators, and install hydronic baseboard heating in its place.

EARLY BIG PROJECTS – Our back yard was level for about 30 feet behind the house, then rose rather abruptly to an elevation about five feet higher. **Making a patio** entailed installing a stone wall to meet that rise, forming an elevated flower garden beyond it. I discovered that our back yard largely comprised glacial till from an ancient esker. The topsoil was only a few inches, and below that was a mixture of sand and rocks—lots of rocks. I built a sifter to separate the rocks from the sand, thereby obtaining most of the sand needed for the patio base.

Two and a half tons of Pomfret flats for the stonework were delivered. Some of these were so large that I could barely move them. Eventually, the wall was finished to about two feet above the future patio surface and about twenty feet long. After finishing the wall, I picked up a carton of stone chips and pulled my back quite badly. How ironic that I had been so careful moving those tons and tons of large stones without incident, and now I'd injured my back by lifting tiny scraps. But not to worry—my children were employed to lay the patio. On my hands and knees, I leveled the sand with a long straight board while the 8-, 9-, and 10-year olds fetched the bricks by wagon from the pallets in the driveway and placed them in a herringbone pattern. They did a beautiful job and the patio served us for decades.

The **steam boiler** must have been original with the 1924 building of the house and was huge, taking up lots of space where my shop should be. Also, it was very inefficient as evidenced by the sweltering heat in the basement all winter. However, there was also a tiny boiler that had been installed 15 years earlier for heating the addition. I attached an electric clock to its pump circuit and determined that the duty factor of the smaller boiler was very low—only a few hours a day, the most being seven out of 24 hours during a severe cold spell. I reasoned that it had plenty of capacity to heat the other half of the house. And, it did.

With much effort and three-foot pipe wrenches, I removed the pipes, radiators, and boiler. Once the radiators were out, the plumber installed baseboard convectors connected to the newer boiler.

That newer gas-fired boiler had a wonderful feature that we used during Wayland's frequent power outages. If I leaned a 2x4 against the control relay, and manually opened the zone valves, we could keep the house heated throughout a multiday power outage via convection alone.

• • •

The most extensive modifications I made to the house were the addition of a great room and the remodeling of the kitchen in 1986.

That project was postponed for about seven years while we figured out a good solution to the basement stairs. Even as we were buying the house in 1979, we intended to build a great room extending to the rear from the extant kitchen, but the only stairs to the basement would end up in the middle of the new addition. We wrestled with various simple solutions, but ultimately, my wife convinced me that relocating the stairway to be under the second floor stairs was the only solution despite its difficulty.

THE ADDITION – Moving the basement stairs would affect the heating pipes, the gas line, the central vacuum system, the main drain pipe, hot water, cold water, the video cable, and the only access doorway between two halves of the basement. Such an undertaking had been too much to envision, but after getting reasonable quotes for doing all that, I contracted a concrete-cutting company to cut a new doorway through the foundation and hired a plumber to move all those pipes. The project could begin.

One day in May I built the new stairway to the basement. Next we contracted for digging and pouring the foundation without damage to the nearby brick patio that the kids had helped build years earlier.

On Friday, July 4, 1986, I began demolition of the little 7 by 7 foot rear entryway that included those troublesome stairs. Soon, it was gone, and the adjacent shingles and siding removed enough to begin construction of the 12 by 18 foot addition.[18] I was taking the next two weeks off, and the rest of the family was going to Cape Cod. A huge stack of all the wood needed for the entire project was already in the driveway as was the steel beam needed to support the second floor after removing 18 feet of the back wall.

First, the floor framing and decking was built. Next, I dragged and rolled the 12-inch steel channel beam, up onto the deck where I could drill all the needed holes. I selected the best locations and drilled corresponding holes through the beam. Next I constructed a temporary guide to capture the beam while lifting it into place and prepared props for intermediate and final positions. The already arranged "beam team" comprising eight neighbors and colleagues from work showed up after their work day on the appointed afternoon, and working together lifted the beam into place in a matter of minutes. I placed the props, added a few clamps for good measure, and we all shared in the beer supply I had laid-in for the occasion.

After figuring what the approximate load on that beam would be by estimating the weight of roof, shingles, attic flooring, second story exterior wall, floor, bath fixtures, wallboard, tile, *etc.*, I calculated that the expected deflection of the beam would be $5/16$ of an inch. The next day, I fine-tuned the beam's position, outfitted it with a house jack at each end, and began the attachment process at the center: first three or four carriage bolts near the center, then jack the two ends by $1/16$ inch, then one more bolt to each side of the first group, then a little more jacking. This process was repeated until the beam ends had been lifted by the full $5/16$ inch, all two dozen bolts installed, and a stud to the deck glued and nailed in place. I would not know for a few more months if my estimate was correct.

Next came the walls, which I simply framed on the floor and lifted into place. The rafter trusses were similarly framed on the deck. The ceiling joist was a 22-foot long 2x8, and after the first truss was measured, cut, and nailed, I traced it onto the floor with magic marker to speed up the making of the other eight. Lifting those trusses onto the walls was possible only one end at a time. After the first four were

---

18 There is a side story associated with demolishing that little extension. I started work about 9 am. At about 3 pm, I was up on the roof prying boards loose when I realized that I was so exhausted that I was actually sliding my hammer and pry-bar across the roof instead of lifting them. I came down to rest. I was depressed...only six hours of work and I was worn out...there were still hundreds of hours to go...what would I do? While resting, I got a glass of water and drank it. It tasted good, so I had another. Within thirty minutes, I had seven glasses of water, felt recovered, and went back to work until darkness. It suddenly struck me: At no time did I ever sense thirst, but apparently I was grossly dehydrated. Moreover, I realized I was having the same sensation I had in 1964 in the final stages of climbing out of the Grand Canyon. I ran out of water both times and never knew it.

in place but hanging upside down, I rotated them upright one at a time, adjusted their position, and temporarily kept them upright by tacking on a piece of strapping near the peak. Then I used a stepladder to climb up between two of them to mark the line of intersection with the house. While still between two trusses, I managed to dislodge the one nail that was holding everything in place, and all four of them came crashing down, the ladder skidded out from under me, and I found myself hanging by my armpits from the now-flatted trusses. I was also being pinched across my chest, and it took a bit of squirming to dislodge myself and drop down to the floor. Next day, I had huge hematomas under each arm.

The rest went smoothly but was time-consuming: plywood roof sheathing, asphalt roof shingles, wall sheathing, sliding doors, cedar shingles, rough wiring by an electrician, insulation, 12-foot sheetrock onto the ceiling with a custom rig of my design, and finally wall sheetrock. Now I was ready to shut down our kitchen and remove the original exterior wall.

Kitchen operations were moved to our extra kitchen—the mudroom. We depended on the toaster, electric fry pan, and microwave for cooking for the next six weeks. I removed the old appliances, cabinets, and a window. When I sawed through the studs of that originally-supporting exterior wall, the weight from above did not even pinch the saw and the measurement from floor to beam did not perceptibly change. That $5/16$ inch of predeflecting the beam was perfect.

The next six weeks were intense. The old kitchen ceiling came down; the door to the living room closed up; a new door opened into the foyer; wallboard removed; linoleum, plywood, and maple flooring cut and crow-barred up in small pieces; the new and old floor knit together with final interleaved sheets of plywood; plumbing for the bathroom above rearranged by a plumber; metal-stud ceiling joists installed to cover plumbing and all those original multiple levels of ceiling; a box constructed at the beam; more can lights, switches, and outlets contracted; sheetrock installed, jointed, and painted; new oak flooring installation and sanding contracted out; flooring varnished; casing and baseboards installed; cabinets brought in from the garage, leveled, aligned, and installed; laminate tops measured, ordered, and placed; sink, dishwasher, range, microwave, and refrigerator installed. All that in six weeks along with my day job. And the only thing that prevented us from fully using the kitchen for Thanksgiving was that the oven had a defective thermostat.

The following spring, I built a deck around all three sides of this addition for egress through the sliders. Two sides about 6 feet wide, one side 12 by 18 for table, chairs, and gas grill.

The only sizable project after that was the remodeling of the master bath a decade later with my son's help. Everything came out, and new versions went back in. It followed virtually the same plan but had a number of subtle improvements including water-resistant tile-work, a jackable support for the cantilevered counter to avoid the previous sagging, and more cabinet volume.

During the 29 years we occupied that house, it seems like we were frequently undertaking some project including extensive landscaping. In preparation for selling the place, I had to shape matching shiplap siding for replacing a few rotted pieces on the stable since that '30s style siding was obsolete. And over the years, every room was redecorated, some twice, and a few included significant repairs or changes. My wife did most of the painting and all of the wallpapering—even those troublesome wallpapers that insisted on peeling as fast as they were put up.

In our last decade in that house when we were going to the Cape frequently for long weekends and in anticipation of heading to Florida for months, I built a control system to operate lights. It ended up working so well that I just left it to operate at all times until the house was sold. The system used a Tattletale—one of the microcomputers I had beta-tested and used for GTE, Sylvania, and several consulting projects. That computer could be accessed remotely by laptop modem, had a very low-power sleep mode during power outages, and had more than enough driving outputs. It not only controlled lights, but also collected both interior and exterior temperatures and kept track of power outages.

I configured the program to turn on lamps in the living room, family room, my office, the kitchen, and upstairs bedroom in patterns that we would ordinarily use. The front porch lights came on from sunset to 11, and the driveway spotlights only some evenings. All of these times were randomized with an algorithm to vary both on- and off-times by up to 30 minutes. The program included a sunset calculator for that porch light and the spotlights. It also included a method for overriding any one of the lights either on or off by remote modem, or locally with a pushbutton switch in the front closet.

A Nearby Remodel – In 1987, the year after remodeling our kitchen in Wayland, I had become exasperated with the Cambridge apartment house. The recent tenants were not acquaintances, they were calling with complaints that needed long trips to resolve, and the Rent Control Board was unreasonable about every request for fiscal relief. I sold that building and bought a single-family diagonally across the street because I had had so much fun remodeling ours the year before. I wanted a similar project but it turned out to be an albatross.

I hired a realtor to evaluate the eventual asking price were I to implement all of my planned updates. The plan looked good and the profit tidy. Soon after the purchase, the realty market tanked and I was looking at a problem. I chose to cut back on contracting much of the work and doing more of it myself. But doing things myself slowed down the project, resulting in much more interest cost because the low starter-rate loan was rising every six months. It eventually took over a year to finish everything.

IMPROVEMENTS – I hired my son and his friend to repaint the house a dark gray with white trim—looked much better. They also did some demolition work: tearing out the two bathrooms, demolishing the kitchen, and taking down the ceiling and wall in the kitchen-family room area. But once they went back to school, progress slowed to the pace I could manage myself with periodic help from a carpenter.

A master bath was added at the end of the house. For that, I dug for and contracted the foundation construction and hired the carpenter to build and roof the structure. The carpenter also converted the miserable room above the garage into a decent space by combining it with the kitchen, forming a huge cathedral ceiling over both, and replacing the little side door with a 6 by 8-foot swinging patio door with a half round transom above. I designed the kitchen area, ordered and installed custom cabinets, and made a plywood template for a huge, U-shaped countertop. This part of the house was a lot of work, but it was magnificent when done.

The two double-hung windows in the dining room were replaced with a French door leading to a small newly-built deck.

When all these spaces had been reworked, I hired a plasterer who could fit in the job because it was so "small." It would have taken me two months just to hang all that sheetrock and another month or more to tape it all. This crew put up all the sheetrock one day and plastered it the next. I spray-painted all the woodwork including the newly-replaced paneled doors, and roll-painted the walls and ceilings. The bath floors and showers were professionally tiled, and I installed the cabinets.

It took over four years to cycle that house. It was rented twice, once to a family between houses, and once for a business school family sent east by his company. Also, my 50th birthday party was thrown in this empty space—a sock hop to spare the newly-varnished floors. Overall, that project included a lot of agony. Rebuilding someone else's home was far less satisfying than doing our own. It was not a financial success for us, but mixed in with all the other real estate holdings it was just a hiccup.

Eastman, New Hampshire – Also in 1987, we purchased a vacation house in the Eastman community in New Hampshire. Originally, we got it as a place just to get away, but during our first winter of visiting there, we happened upon an offer for a "B-Day" (beginners day) at King Ridge Ski area only a few miles away. The offer was reasonable, we thought it might be fun, and since our son was six it seemed time to introduce him to the sport. That turned out to be a terrific introduction to an exciting activity we kept up for years.

The house was furnished and had three floors on a small footprint: three bedrooms, full bath and laundry on the first floor; one large space for living room, dining, and kitchen on the second; and master bedroom, bath, and balcony at the cathedral ceiling on the third. The second floor also had a screened porch and the third level was within the steeply gabled roof.

This was supposed to be a place for relaxing, but as usual, I undertook several projects. The first was installing a roof antenna with rotor to hopefully improve TV reception over the rabbit ears. That required the purchase of a 40-foot ladder. It was also a dangerous project—all alone with no cell phone on hand. And reception still bad, so we signed up for cable service.

MAINTENANCE – I did, however, have another use for that long ladder. There was a roof leak that a roofer tried twice to fix without success and refused to return a third time. I put the ladder up, figured out why it leaked, and fixed it.

Taking on complex or detailed construction projects was not so easy given that most of my tools were two hours away. So when it came time to build a proper front stoop, I built it in Wayland in the convenience of my shop and with appropriate tools. Originally, there was a single large timber used as a front step, but it was narrow and resulted in two unusually high steps. I designed a small landing about three and a half feet square with a diagonal step cut across one corner. When finished, I put it into the station wagon, took it to Eastman, dug post holes for the corner posts, set it in place, leveled it, and backfilled. Voila! Two steps, adequate tread width, and the upper step formed a generous platform to stand just six inches below the door.

But the biggest project was making the water easy to drain. We soon discovered that the cost of keeping the house from freezing with electric heat during a mere week away was astronomical. So I tuned up the water system so that it could be quickly drained, the traps filled with antifreeze, and the heat turned off at the end of a ski weekend. This required four major changes: easy attachment of a blow-out compressor, solenoid valves in the crawl space for dumping the last of the water, a catch-pan under the water heater with a drain to the septic tank, and applying heating tape to two horizontal pipe runs in the crawl space.

In practice, as winter approached and I shut down the house water for the first time, I used the compressor to blow out the lines to the dishwasher and washing machine. Thereafter we did not use those two machines during weekend visits. At the end of a weekend, I first switched off the water supply, then followed a sequential checklist that included opening each hot and cold valve to let the water drain backward into the first floor bathtub. After opening the drain bib on the water heater, I carried around a gallon jug of antifreeze and used a measuring cup for adding a specific amount to each trap and toilet tank. Lastly, I switched open the solenoid drain valves in the crawl space and turned off the heat. All this took about ten minutes, and the water heater was still draining as we departed. Of course the house was incredibly cold when we returned the next weekend. We usually flipped the thermostats on and went out for supper—but it was still quite cool when we crawled under our electric blankets.

The house was a delightful place for us on weekends both winter and summer. We invited many guests there, and our dog Sam must have loved it also—whenever we began to pack the car, Sam became excited, jumped willingly into the car, slept most of the way but always awakened as we approached the exit from the Interstate. We have many happy memories of those days, but eventually we all outgrew King Ridge, found ourselves going there less often, and decided to sell the house in 1993. It was a good investment.

Vacation House, Cape Cod – Some years after selling the Eastman house, we thought it time to find another vacation home, but on Cape Cod, not in the mountains. We looked in several towns without success until my wife took an exit from the highway just to shop in a needlepoint store. While driving through the town in February 1999, she spotted a small *Open House* sign, braked the car, and turned into the street to see it. We liked what we saw and the price was right, so we thought seriously about it for the next week.

The house was being sold by an eccentric professor who made the process so difficult that we almost abandoned the effort. Within the mere two weeks we negotiated with him, he took the house off the market, put it back on under a realtor, and then withdrew it from the realtor. When we eventually agreed upon the purchase, he wanted us to pass papers within two weeks then rent it back to him at ten dollars a month for six months. Even after we dismissed that foolishness, he still wanted to rent the basement indefinitely for his enormous book collection. We made sure everything was out before we went to the Registry of Deeds.

Before moving in we had the entire interior painted, the floors refinished, two rooms carpeted, and I installed new bathroom sinks, counters, vanities, and lights.

We visited there many weekends that first summer, had the kids join us a few times, and spent one full week of vacation there. The house was in good shape, but much of the yard had gone into brambles. We were reluctant to remove those brambles because of the increased care that any new lawn or garden would require. But I did cut the vines back from the house and out of trees, clear a few pathways through the yard, and spray a lot of poison ivy.

The front lawn was the richest I had ever owned and wanted to keep it that way, so over the summer, I installed two underground sprinkler heads and rigged up a manifold at the front faucet for automatic lawn watering.

The house was in good enough shape that we could enjoy our weekends and week-long visits there without having to work all the time. But there were a few larger projects that I undertook over the years while working there alone. The kitchen was one of those. Most years in spring or early summer, I would take one day off from work during the week and go there just to weed, trim bushes, prune trees, and pull out vines. It was a long day that left me exhausted, but it freed up my visiting times from having to constantly putter about the yard. Another project was to slowly replace the shabby and rotting window sashes with insulated sashes and spring balances. Also, I designed and specified plumbing changes to eliminate the long wait and the scald problem in the showers. (Also see *Plumbing Orthogonality* in the *Adult Experiences, Tales and Anecdotes* chapter.)

PROJECTS – A remodeled kitchen was the only major project undertaken at that vacation house, and that was done in 2000 before our second summer. I designed the new one over the winter and installed it the following spring. My routine was to spend a three-day weekend alone and work long days. I ripped out the old kitchen, rearranged some wiring, put in a new Anderson window, repaired the walls, and had a new oak floor installed. To bring in the bulky cabinets alone, I made sturdy custom ramps to fit the garage and deck steps. One challenge was that the nearest table saw for making special shapes was 90 miles away but with detailed notes I was able to make the parts during the intervening week. All the cabinets were installed over two weekends. When the owner of the Corian shop was making the counter template, he abruptly stopped at one point, stood back to survey the cabinets, and asked, "Did you install these yourself?" I replied hesitantly, "Yes, what's wrong?" "Oh, nothing's wrong, but the professionals don't usually get them this straight." A nice compliment.

We wanted a way to turn the heat on well ahead of visiting for a winter weekend, so I designed a control system using one of the miniature Tattletale 5F microcomputers I used at GTE. I designed it to be accessible by phone with my laptop's modem, to monitor inside and outside temperatures, and to turn up the heat on request. When we departed from the Cape, we turned the thermostats down to 45° for the week. Prior to returning to the Cape, I used a laptop to call the Cape house, hang up after four rings to avoid reaching the answering machine, then redial for another four rings. The Tattletale would pick up after a total of eight rings. With a few keystrokes, I would set the house to heat up via a relay to bypass the thermostat for 24 hours and regulate to 70°. If we did not arrive for 24 hours, the house would cool down again to the thermostat setting. If we did arrive, we reset the thermostats to a comfortable level for the remainder of the weekend.

A monumental bonus for having this system occurred a few days before departing for Florida in the winter of 2005 when the microcomputer informed me that the house was getting cold. I looked at the data through my laptop and observed the interior temp to be 36°F and the outside temp to be 109°. In January? Obviously, the temperature measuring system had gone haywire. I tried to ignore it, but could not sleep. Finally, at 3:00 am, I got up, drove to the Cape, and discovered the house to be 30°. The pilot light in the boiler was out. I relit the pilot, drained the water from all pipes except for the heating system, and put antifreeze in all the traps. The heat was left on, but if it were to fail again, no water damage should occur in the household plumbing—only, perhaps, in the boiler's heating pipes.

Soon after I got there I flushed the toilets to avoid possible freezing. The water must have been supercooled because as it was flushed, I noticed ice rapidly forming in the bowl during the flush. But the real surprise was upon opening the tank to add antifreeze. Apparently, as the water began to drain from the tank it formed a thin sheet of ice over the surface. But since the surface was moving downward the ice was formed in an intricate, delicate, three-dimensional pattern throughout the tank. It was both fascinating and beautiful.

Ultimately I found that the outside thermistor had actually failed, and that the failure was within a few hours of the boiler's pilot light going out. So much for the engineering adage that two unrelated failures never occur simultaneously.

Soon after my second and final retirement, we began to talk of moving from Wayland to that house and discussed what it would take for it to be livable for us. We talked to builders about bumping out here or up there, but those ideas were inadequate, expensive, or unworkable. We got a price for finishing that relatively sunny, walk-out basement into a large game room with billiard table, an office, a shop, laundry room, and full bath. With pull-out sofas, that would give us two additional sleeping areas. To give us storage for our many projects and interests, I spent several weekends reversing the basement steps and building real stairs to the spacious and floored attic. We even designed and built a model of an extensive addition and hired an architect to refine it. In the end, all of this was to no avail primarily because the lot was too small for these changes and the septic system they would require.

We were both discouraged, but in the fall of 2006 my wife saw a realty ad on the Internet for a house in Osterville that appealed to us both. Within a week we made a bid on the house.

Our Home in Osterville – It was several months after that first viewing when we discovered that both of us had immediately liked this house. I was only nine feet inside the front door when a very positive feeling came over me. I could see the handsome wood floor, the generous living room, a sizable dining room beyond that, part of an offset family room ahead of me, and a sunroom beyond that. There was a distinct feeling of spaciousness that the first Cape house always lacked. Except for the mirror image, the added family room, and the greater size of every room, it was very similar to the other Cape house.

We didn't move in for a year and a half. In that time, we planned and executed several changes and improvements, most of which were done by contractors. We also spent two winters in Florida. The first project was to redesign the master suite, which comprised a huge 22-foot long bedroom, a tiny, shallow closet, and a narrow one-sink bath. After our Florida winter, a carpenter was hired to reconstruct it, moving the eastern wall four feet west, enlarging the closet to become a walk-in, and lengthening the bath enough to have two well-spaced sinks. The bedroom remained generous at 18 by 13 feet. The carpenter ran out of time so I ended up installing the sheetrock, finishing the joints, modifying and installing the mirrored sliding closet door, casing the doors, finishing the closet, constructing the shower pan, and ultimately tiling the entire bath.

We hired a painter to strip wallpaper and paint the entire interior a lighter color. Then the oak floors were refinished and the quarry tile floor of the sunroom replaced with matching oak. All of these projects were done in the spring, summer, and early fall.

At this time, we had no pressure to move as we were still trying to get the Wayland house organized, cleared of clutter, and prepared for sale. Also, we were spending four months in Florida with a *what, me worry?* attitude.

During the above projects, I went there often to oversee them or do some landscaping since we wanted to get the lawn improved and new shrubs growing as early as possible. A friend did a plan for our front yard. By repairing the irrigation system, patiently digging out the dandelions, fertilizing, and occasionally seeding it, the front lawn eventually became verdant.

We also researched swimming pools that season. I surveyed the back and side yards and added the measurements to the house dimensions already in my CAD system in preparation for an accurate plan of action. Over the following winter in Florida, I planned the pool project and contracted installation of a one-piece fiberglass pool.

In the spring of 2008, we had some rhododendrons transplanted temporarily, trees removed, and the back yard excavated for the 30 by 13-foot pool. The pool was lifted over the house by crane and set in place. Then I attended to the final touches of contouring the soil and marking the outline of the concrete pool deck, which was poured and textured. Next came the construction of the waterfall by a mason-artist. The rhododendrons were replanted and three more huge rhodys brought in. I did the final grading of the topsoil, then had a fence installed.

Hiring out a pool construction job should have been easy on me, but it took a lot of my time and work. I measured the back yard, made a full-sized polyethylene template of the pool, and tried it out in several positions. The telephone cable that was found to cross the back yard had to be shifted safely out of the way without cutting; I devised and installed drainage lines for both pool deck run-off and future connections to the gutters; constructed a retaining wall around the filter system; and staked out the fence trajectory. The pouring of the pool deck didn't need me.

We moved into the house in June 2008 and began using the pool in midsummer even before the new rhododendrons were planted and the final fence was installed.

LITTLE PROJECTS – Also that summer, I built a low brick wall for accenting our side door entrance. I designed the L-shaped wall, excavated and poured a foundation, and selected the bricks. The pros make it look so easy, but I found laying bricks to be arduous. It took me over two weeks just for the six-course wall construction. It is beautiful and has not cracked through many freezing winters. I reset the enclosed loose-brick patio area at twice the original size.

The next summer, we had the deck rebuilt to our design. By plan, I finished the railing myself with custom fittings and special touches to accommodate tempered glass: $3/8$ thick, with a gap at floor level. This design optimized our view of the pool area when sitting on the deck. I also built a solid, shingled end-panel to give us more privacy from the west.

In 2010, I rebuilt the kitchen. We lived with the original kitchen in Osterville for two years, not because we didn't know what we wanted, but because it was satisfactory and other projects like the pool and deck needed more attention. While in Florida, I finished the detailed plan and ordered the cabinets.

BIG PROJECT – About a week after our return, the project began with a day of clearing out the cabinets and another day of demolition. Then I patched the ceiling and walls where lights and cabinets had been removed, the electrician put in new outlets and newly-located ceiling lights, and the painter painted the ceiling, which included the family room, side hall, front hall, and bedroom hall.

While planning such projects as this, it is so easy to forget details. For example, to paint the ceiling to *my* satisfaction it was necessary to drop or remove fixtures instead of just painting around them. Including lights, A/C ventilators, chimes, alarms, smoke and CO detectors, there were 24 items that needed some action and parts stored in an orderly manner. That was my job, took a lot of unplanned-for time, and not everything cooperated to be put back.

Getting that much done took twelve days including time off for selecting new appliances, breaking for a Mother's Day party with family, and selling the old cabinets and appliances through the Internet.

After painting came the floor. This began a period of impatience as there was virtually nothing I could do for almost a week. Our food, coffee, and paper plates were stored in the bathroom, and we had to walk through the front yard to get to the refrigerator or microwave in the garage. We ate out a lot. The floor man wove the new kitchen oak flooring into the original family room boards with expert invisibility, and three coats of varnish were applied before beginning installation of the new base cabinets. Plastic drapes were taped over every doorway, and it took a lot of patience to watch the varnish dry.

The cabinet installation went smoothly and was finished with lots of shims neatly glued in place. I had built several specialty ramps and levering fixtures for moving the cabinets from the garage and into place without scratching the new floor. It took equally long for me to build the complex structure for the backside of the peninsula and its elevated "bar-top." Thus another twelve days had passed.

Next a template was made for the countertop, followed by another waiting period. But I used this time to install new floor tiles in the concurrent project in the laundry and powder rooms. The floor man came back for one final coat of varnish, and this final coat extended to include the entire integral family room so that the flooring joint was truly invisible. So another two days for varnish to dry and we could finally purge the house of plastic drapes, blue tape, stacked-up furniture, and dust galore. Thus another ten days passed.

Soon the Corian countertop was installed. Within a few days of that, I plumbed the sink, installed the dishwasher, disposer, refrigerator, cabinet fronts, and drawers. I also finished grouting the tiles in the powder and laundry rooms and reinstalled the washer and dryer. Now we could live again.

At this point, thirty six days had elapsed, and we had a "waist-high kitchen" and full use of the house—except for those missing wall cabinets, the microwave in the garage, and no toilet in the powder room.

After a few days off while sniveling through a cold, the wall cabinets and the microwave were installed; some trim completed in the kitchen, laundry, and powder room; and the toilet reinstalled. We started on May 10 with clearing out the cabinets, and I considered the kitchen effectively finished 55 days later on July 4 when the last wall cabinet was secured in place. There were, however, a few aesthetic details here and there to be finished.

Self-critique #1: After 42 years of doing this stuff, I am still too optimistic about how fast things can be done.

Self-critique #2: Now that I have become more mature, and therefore more patient, projects proceed much more smoothly."

• • •

One final project in this house was the 9-foot sliding patio door from the family room to the sunroom. This door was tired and troublesome, and we wanted one with all three panels movable. But the only interesting part of this story was the removal of the old door. So here it is.

After considering other approaches, I decided to break the glass and remove the doors in buckets and broken pieces of frame. I laid a large drop cloth and a sheet of plywood on each side to protect the floors and make it easier to shovel up the glass. Wearing safety glasses, a face shield, long sleeved sweatshirt, and gloves, I placed the tip of an automatic center punch against the glass, and pushed. Presto. The entire glass instantly formed a spider web of cracks but didn't fall out. Even after 45 minutes of periodic cracking, I needed a stick to clear away the broken glass. I shoveled up a huge bucket-full of glass beads despite the glass being only $1/6$ of an inch thick. When all the glass was cleared, I sawed through the side stiles and removed the frame in pieces.

Out of curiosity, I tried breaking one glass panel with a hammer. No way! I swung the hammer repeatedly, building up to a resounding crescendo and hitting harder than needed to drive a 20d nail. It wouldn't budge. So if you want to escape from a sinking car with electric windows, you better have an automatic center punch in your glove compartment.

• • •

Ultimately, we found the trek off the Cape, up Route 3, and through Boston to visit our Massachusetts children and grandchildren increasingly difficult. Apparently our children did too, as their visits became less frequent. Also, we found caring for a house and one-acre yard too challenging especially in light of our winters spent in Florida. We began to look north of Boston.

Our Condominium in Naples, Florida – A few months after buying the Osterville house, we bought a winter condominium in Florida. Even living in a condo, where one is supposed to relax and let others do all the work, I couldn't help but take on projects both inside and out. We purchased the condo in March 2007 and immediately contracted for it to be painted, retiled, and recarpeted. We returned in midsummer to bring some furniture, inspect the painting, and oversee the installation of tile and carpet.

When we first occupied it in January 2008, there were many minor projects started like tracing the electrical circuits and installing shelves. But larger projects included selecting and installing a new kitchen sink, faucet, and disposer; installing five recessed lights in the kitchen ceiling; and constructing valences for the dining area using foam-core and poster board!

Outside projects included many taken on for the community. We weren't there long when my wife volunteered me to serve on the gate committee. I ended up surveying the entrance and making a CAD drawing for choosing the best site for the gate lifters and diagrams for the construction permit. That project continues to dog me as I have informally become the point man for it. I can modify the gate-opening codes from my laptop and operate the gate from my cell phone. Useful for open houses.

CONDO BOARD SERVICE – Soon I was asked to join the Board of Directors and was appointed to a vacant position in February 2008. They asked me to evaluate the efficacy of replacing the propane pool heaters with geothermal heaters. I worked through half a dozen scenarios using Excel to estimate the savings. It surely appeared that geothermal heaters could save us a lot of money. The Board decided to have wells drilled down to the aquifer and three heaters installed, one for the spa, two for the pool. A year later, I confirmed that we had saved about $10,000 in fuel over twelve months, and that included pool-cooling in the summer months—a bonus the neighborhood had never previously enjoyed.

After a year on the Board, I was elected president in February 2009. It has kept me busy, and I have enjoyed it. In the spring of 2011, the Board undertook the matter of repainting the buildings. Seven years earlier, the same project had been very contentious, and owners greatly disagreed on choice of color. I managed to have the Board move delicately and to keep the owners well informed of progress, including an owner-vote to select the final colors.

After three years, we moved across the street into an identical condo, but one with a sunny breakfast on the lanai, cocktails in the shade, and the view and serenity of a lake, a golf course, a distant fountain, and a few hidden houses far away.

We decorated this condo identically to the previous unit, which meant repeating many of the projects done three years earlier. Even the pictures were hung within a quarter of an inch of their original position. We like the new place, especially the typical silence even with the sliders fully open. And we love our year-round summer life.

Our Home in Newbury – In 2015, we looked for and bought a condo in a recently-developed community in Newbury, Massachusetts. I had expected to have a very hard time leaving that Osterville home with all the custom details we had found or added to it, but the ease of living in a condo and the proximity to children negated that downside.

The recently-built condo is very nice with large spaces, master suite on the first floor, and guest areas on the second. Except for building many storage shelves in the 9-foot high basement, reproducing my shop, and putting higher counters in the master bath, it hasn't needed much attention. Thus there is little to report.

Remodeled Kitchens – My son asked me how many kitchens I had remodeled including the just-finished one in Osterville. I mentally worked my way through the list: There were two in Cambridge; the new one upstairs and the remodeled one in Belmont; two major rebuilds in Wayland, one in our own home and one at the investment house across the street; and two more on Cape Cod. Eight in all. He gave me a plaque to commemorate them.

In retrospect, I found remodeling kitchens to be very satisfying. The many stages of making over a kitchen include my childhood interest in architecture; my everlasting interest in the design process; the detailed planning and specifying of components to be purchased; working with my hands as the kitchen gets gutted and rebuilt; designing for and working with plumbers and electricians; the inevitable woodworking to make everything fit; the aesthetic touches to be applied at the end; and the daily enjoyment of the result both by myself, my wife, and typically many others who visit.

Remodeled Baths – As long as I am recounting kitchens, I should recount bathrooms, especially since there were even more baths than kitchens. Some of these were described in other sections.

> **BATH COUNT** – The **zeroth** bath was the shower room in our childhood basement that my brother and I initiated but made minimal contributions. The **first**, in my Cambridge apartment in 1969, was undoubtedly the most extensive that I undertook. The **second** and **third** were also complete rebuilds at that apartment house, but I had the help of the tenants for some of the work.
>
> The **fourth** bath remodeled in 1976 was the master bath in Belmont. The **fifth** was the first floor half-bath, originally a small, drab closet with toilet only in which I added a narrow countertop with an unusually narrow and triangular sink bowl; added wainscoting, wallpaper, a giant mirror, and a new toilet. Thus it became a decent powder room. Its most challenging detail was coaxing the plumber to cut the sink drain into the old cast iron sewer pipe.
>
> The **sixth** was a full bath I mostly built during a weekend trip to my parents' home in Charlotte in the mid 1980s. My father in his inimitable indifference to the hardships of fellow family members chose to do nothing about the miserable first floor half-bath where Mother was restricted because of her arthritis. After taking copious measurements during one of my visits, I worked out a design for expediency, had key components delivered, and hired several contractors to show up on a three-day weekend during which we demolished two walls, built four new walls, installed a four-piece shower, blue-boarded everything, turned the toilet 90 degrees, installed a corner sink, wired-in new lighting and a heat lamp, patched the floor, and installed a wall-to-wall loose carpet. The following Monday after I returned home, the plumber hooked up the sink, toilet, and shower. Later, my brother, Bill finished the door casing, and the walls were jointed and painted. The major challenge to this job was having everything planned for in advance and on hand when I arrived.
>
> The **seventh** and **eighth** bathrooms were the two at the investment house across the street from us in Wayland. The **ninth** bathroom was the remodeling of our master bath in Wayland with my son's help. The **tenth** was the master bath we resized and completely rebuilt in 2007 at our home in Osterville.

Perhaps there is an **eleventh** bath if a partial counts. At our current northern condo I designed and installed new vanity cabinetry for more useful storage and a two-level sink-top to serve our different heights.

CONSULTING

Early Consulting – My first experience with consulting outside of my day job occurred in the mid 1960s not long after joining Polaroid. Several of my fellow grad students at MIT had formed a company, Icon Inc., with a specialty that included optics and control systems. One of their clients needed a motor-driven method of changing tools on a lathe for the automation system that Icon was developing for them. I was hired to design a six-station tool turret system that was indexable one station at a time with a motor. I designed it for them and got paid for my first consultancy.

In 1970 to my surprise, the research doctor at Mass General Hospital for whom I had designed my bachelor's thesis in 1962, asked me to design a larger version of it—the optical model of an ultrasound cone. Upon studying the geometry, I found that the original focusing mirror could be reused for a larger cone, but it would require a larger housing. The cannibalizing of the original unit was agreeable to the researcher, so I laid out the new design, detailed the individual parts, and delivered the drawings to Mass General. I was paid, but never saw the resulting hardware.

• • •

In the fall of 1979, four years after I left ECA, a fellow manager while I was there, called me to say that he had left ECA, formed his own company, JB Systems, Inc., and needed some design help from me. He, too, was in the flame-monitoring business but felt he had a better product than ECA. I designed a pipe-mounted housing with a self-checking shutter system that mounted directly on the circuit board. I also contracted a shop to build prototype parts. It was a clean design with all the exterior parts of gold passivated aluminum. An important light baffle was made from brass using chemical milling and included all the needed fold lines and his logo partially etched through the front panel. That was the first of several occasions for which I specified chemical milling for making parts. He was delighted with the logo.

I designed a second product for him that included a new challenge for me. This product was to be built in England for a European market, and they wanted the drawings to be metric and to be drawn with first-angle projection, something I had never done. To clarify: In the US, the front face of a part is shown at the center of the drawing, the right face is projected to the right, the top face is drawn above, *etc.*, and this convention is called third-angle projection. In Europe, the right face is drawn to the left as though the part were tumbled to the left to reveal the right face. Similarly the upper face was drawn below and the lower face drawn above. There certainly was a degree of logic to that geometry, but it was surprisingly hard for me to draw that way after so many years of projecting the other way. Designing in metric was not difficult except for finding the standards for screw thread, head, and nut dimensions.

Avant Inc. – In 1980, I was recommended to Avant by a friend for some minor debugging work that turned into a ten-year relationship. Avant was a small company that made identification cameras using Polaroid backs. They also made ancillary products for producing ID cards and badges integrated with a picture of the person. I ended up consulting for over ten years from early 1980 into 1990, designed four distinct cameras, a few accessories, and introduced some innovative systems such as the vacuum-formed light tunnel with chemically-etched dividers.

Their first project for me was to straighten out an unworkable design of a four-image shutter. They had a roughly five-inch square aluminum plate with four lens holes, each outfitted with a shutter blade, a return spring, and a capacitor-fired solenoid to kick the blade. Like its predecessors, this camera had dividers in its light cone so that four small images would be formed onto a single sheet of Polaroid film. The springs weren't doing their job. I changed the design from a long, but still too-stiff tension spring to a torsion spring with a long stroke. I drew up each of the associated parts for them to use for manufacture and generated a parts list for the purchased pieces. The owner was so impressed with these simple drawings that he asked if I would design the rest of the camera; which I did, introducing several innovations—at least innovations for Avant.

That first camera was basically an upgrade of an earlier product, but the innovations decreased the number of parts and simplified others. Avant made most of the parts in their own two-man machine shop without numerical tools, so these simplifications improved their production rate considerably. The most significant innovation was probably the light tunnel—the light-tight cone between lens and film. They had been making it of seven parts: each machined of aluminum, four parts with compound angles, some with angled screw holes, five included light traps to prevent film fogging, six had very narrow slots $1/64$ of an inch wide by 2 to 4 inches long, and it was cumbersome to put together using a dozen screws. I designed a vacuum-formed equivalent of the five-part cone and chemically-etched versions of the two dividers. Both processes were inexpensive, and it took only a few seconds to assemble.

The next camera, called a Roto-Quad, used a rotating turret to move a single lens from one of the four picture positions to the next. Thus, they would save money on the expensive lenses, and the camera would produce pictures for four ID cards while using only one sheet of expensive Polaroid film. All four shutters were fired at the same time, but only the one with a lens produced a picture. A clock motor rotated the turret after each picture was taken and also cocked the single circular shutter blade. I once took this camera into my daughter's fourth grade class, like a show-and-tell, and took pictures of each of the children.

The third camera was a four image camera with four lenses and could be wired to take pictures one, two, or four at a time.

The fourth camera was a six-image camera configured to form six images but hard-wired to take one each of six people, two each of three, three each of two, or six pictures of one person. That camera had very good sales.

Competition for Avant was becoming fierce in my last two years of consulting, the owner was laying off people and consolidating his operation into smaller quarters, and their need for me ended. It was an interesting run.

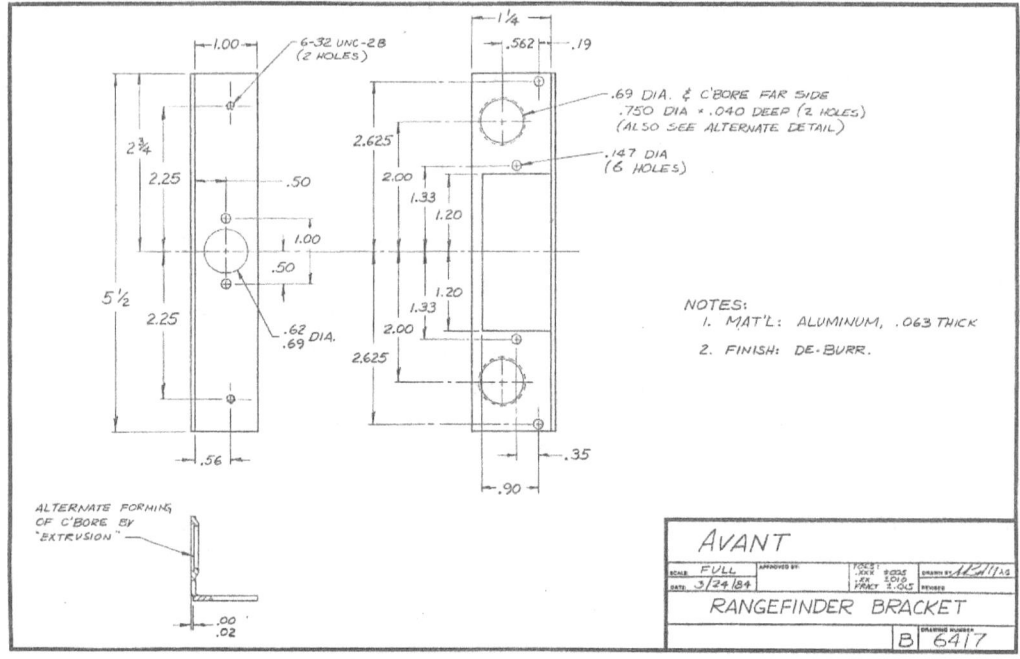

One of many "details" drawn for Avant by hand using a drafting machine-outfitted drafting board, pencil, eraser, erasing shield, and circle templates. Over the years, I must have drawn nearly a thousand such drawings.

Tattletale Consulting – Our success with Onset's Tattletale microcomputers at GTE was well known at Onset Computer Corp. Someone there asked if I wanted to be on their Consultant's List. Sounded like fun, so I volunteered. Over the next several years, I received a number of requests and several projects out of it. For most of these, I never met the user or saw the hardware.

One of the earliest projects came from a forest fire scientist in one of the southwest states. He already had his Tattletale 4A with onboard thermistor and wanted to use it to measure underground temperatures as a controlled fire swept over the area. His fundamental need was simple—just record the temperature at regular intervals along with a time stamp at the beginning and end—and he had already written that much. But his dilemma was selecting the speed. If running too fast, he might miss the fire that may not be set for several days; if too slow, it wouldn't contain enough useful data points during the fire. (This Tattletale model only had 28k of memory.) I wrote a program that ran fast all the time, but once each hour would move the previous hour's worth of data back to the beginning of the memory file if a temperature rise had not yet been detected. If detected, no data transfer would ever take place again. Therefore, there would always be 60 to 120 minutes of data before the fire.

One of the more satisfying projects was the control of a motor that incremented a sampling head down through the mud at the bottom of the ocean. A one-man company in western Connecticut used a submersible vehicle that he designed and built to explore the ocean at relatively shallow depths. One of the instruments was to be settled on the bottom then used to advance a probe into the mud. The control system was needed to advance the probe to selective depths and provide a laptop display to accurately indicate the advancement in fractions of a millimeter. He already had the Tattletale and a poorly working program that needed a rewrite. I wrote a new program, married it to his signal conditioning circuit, and delivered it to his shop where I helped him connect it up and check that it worked. He reported months later that it had performed well on its maiden voyage.

A medical doctor at the University of Chicago was studying the activity level of baboons. I proposed sensing the activity using three axes of accelerometers and analyzing the results into categories of activity level based on intensity, duration, and direction. He liked the plan, but changed his needs many times. I built the circuit board for holding and powering two Analog Devices two-axis accelerometers and integrated it with a Tattletale TFX-11. One day, I tested it on myself by wearing the device tied into a sling under my arm and hidden by my shirt. No one at work noticed the bulge, but the results were fun to study. (After it was delivered, he asked me if I could do it all over again but for a Palm Pilot instead of a microcomputer. I demurred.)

The owner and founder of GeoSystems Analysis in Arizona needed to monitor temperatures at various depths in a blast hole drilled about 150 feet into the earth. He wanted a turnkey instrument including the cable. For that, I purchased a 1,000 foot reel of Category3 telephone cable with six twisted pairs. At his requested intervals, I carefully cut lengthwise into the outer wrap, pulled out one of the color-coded pairs, soldered a thermistor to it, tucked the thermistor back in the cable, and covered the break with liquid electrical tape and shrink wrap. Once it was all assembled, I had to calibrate the thermistor and precision resistor combination. Working with such a big roll of cable presented a bit of a challenge, but the system was eventually delivered. He periodically called me back, first with results from that test, then with vague plans for other instruments, which never came to fruition.

Perhaps the most challenging was for a young North Carolina entrepreneur who was developing a fuel cell to be used as a power source for offshore oceanographic experiments. He wanted to test it at sea while mounted on a buoy with all the rocking and weather conditions that it would actually experience in use. The challenges for me included an unfamiliar Tattletale model with novel features; writing of large data files to a memory board similar to the then-nonexistent but now-common flash drives; and telemetry to the mainland via a wireless modem. I was already familiar with modems, but this one was completely different. It all came together, and my client was happy.

I enjoyed programming. It was a change in pace from my usual work, but hardware and software design shares the same disciplines: robust function, anticipation of all possible conditions, orderly failure and shutdown, compatibility with the environment, *etc.* It also gave me appreciation of the complexity of programs and empathy with my laptop when it doesn't work!

• • •

The most complex Tattletale-related project was done in partnership with Glenn Duchene while we were both still at GTE in 1995. It was for a Resistance Measuring Device (RMD) designed and built for NASA Lewis Research Center. Dave Matthiesen, who had worked on our GaAs Shuttle experiment at GTE and was then working at NASA, recommended us as a possible vendor. The RMD was to be a specialized instrument that could be taken into space aboard the Shuttle and used to measure the resistance of many samples of material processed during a zero-G flight. His expectations of the material properties were so uncertain that the instrument needed to self-range over an enormous scale of possible resistance. The accuracy needed and potential for unpredictable surface resistance required that we have a four-point probe system for these samples that were only 2 mm in diameter. The spring-loaded probes were minute and virtually invisible. Data had to be displayed for immediate evaluation and stored in the instrument for later downloading.

We quoted the finished instrument at $25,000, and split the profits. Glenn designed a circuit that was both voltage-limited and current-limited so that the enormous range of resistance could be accommodated. I programmed a Tattletale 5F to read the measurements, feed back adjustments to the circuit when needed, calculate the resistance, display it with respect to the sample number, and record all data. The instrument had to be secured from floating around within a glove box during use, had an articulating probe head for easy insertion of the samples, a 32-character display, and two buttons for working one's way through the sequence of measurements or through a menu of other options. We delivered the instrument nested in a cushioned case, on time, within budget, and with a complete illustrated instruction manual. We got paid, but at last report, the RMD had not yet gone into orbit.

ABell Engineering – Shortly before taking early retirement from GTE, I formed ABell Engineering with the intent of consulting on my own. Unexpectedly, I took the job at Osram Sylvania and never needed to chase for work but did undertake three mechanical design projects. One company needed a dumpable materials hopper on wheels, for which I supplied several sketches and a cost proposal, but they never responded. I always wondered if they thought my design so bad that they rejected the proposal, or so good that they built the hopper from my sketches. The other two were for a supplier of miscellaneous specialty office items who paid me for two design projects, then discovered that manufacturing new products in low volume wasn't cheap.

Most of the Tattletale programming described above occurred after I formed ABell Engineering. And when my paymaster was transferred from

Sylvania in Massachusetts to Osram in Germany, I could no longer design molds for ceramic arc tubes on Sylvania's office time; therefore, ABell Engineering became a vendor to Sylvania and I continued to design molds on my own time evenings and weekends and got paid for it.

Less than a year after retiring from GTE, I became an expert witness for Fuji Photo Film Company, Ltd. through the Stroock & Stroock & Lavan law firm. That was so lucrative and time-intensive at irregular intervals that I didn't dare take on much more than those small programming jobs that came along from time to time. See *Stroock & Stroock* below.

A Florida law firm looking for an expert witness in camera issues found my name through an Internet search. Their client had a stealth camera for photographing wildlife in the woods that a competitor was badmouthing in their advertisements. I was asked for an independent evaluation of the criticism. My testing found that the client's camera was, indeed, slow in detecting motion and often missed getting a picture of the target. I got paid for my time, but they didn't want a written report or the cameras returned. I gave the cameras to Wayland High School.

An ophthalmologist called on me for suggestions to make a very tiny eye surgery tool. The instrument needed to be like a file but only half a millimeter wide, $1/10$ mm thick, and with teeth $1/40$ mm deep and very sharp. After several iterations, I designed a version with chemically etched teeth on each side, but by this time he had retired and sold his practice and the project went no further.

Stroock & Stroock – In July 1998, Jim DeCarlo, an attorney at Stroock & Stroock & Lavan LLP in New York called to ask if I would serve as an expert witness in a photographic patent infringement case. He described it briefly and said that the job entailed reviewing some claim charts against some single-use cameras, and probably one, maybe two days on the witness stand in court. The client was Fuji Photo Film Company, Ltd.

At the time, I was employed at Osram Sylvania and working four days a week. But from the description, this sounded like a consulting engagement that I could undertake and perform on the weekends and by moving around or consolidating my Friday-off days as might be needed for trips to the law office or court. As the work began to blossom into more than a few hours here and there, I verified that my employment agreement did not preclude me from holding a second job, which included consulting. When the "one, maybe two" days in court turned out to be a two-week schedule, I needed to take the time off without pay.

In the end, I spent eleven years consulting for Stroock, charged for 2,203 hours, took at least five weeks away from OSI work, visited factories in Shenzhen, China near Hong Kong, and testified in Washington and Newark. I dissected and examined hundreds of cameras, wrote 13 expert reports and rebuttals, edited even more witness statements and declarations, endured 65 hours of deposition, and enjoyed nearly 58 hours on a witness stand in eight different hearings or trials. At times it was grueling work especially when the schedule was tight. (Appearances listed below, reports in *Appendix A*.)

My resume must have been near ideal: degrees from MIT; designer of cameras at Polaroid; photographer with historical knowledge of film and cameras; inventor; designer of products, mechanisms, and optical systems; knowledgeable of manufacturing techniques; and familiar with patents as evidenced by the many in my name. (See *Appendix B*.)

INFRINGEMENT – It was very educational learning some of the subtleties of patent law, working with attorneys, and meeting some of the Fuji camera designers. The patents were all protecting single-use cameras, and the early infringers were 23 different companies that collected spent shells and reloaded them for resale. The pending lawsuit was "In The Matter of Certain Lens-Fitted Film Packages, Investigation No. 337-TA-406," which was to be heard by the International Trade Commission (ITC), Washington, D.C. The ITC was being asked to order their import to the US market to cease.

The subtlety of infringement was that Fuji, the complainant, claimed that used camera shells (that Fuji considered "spent," "destroyed," or used up) were being *remanufactured* into distinct new products, and that those products included infringing features. There was no doubt that the patented features were present because the components were all made by Fuji. The respondents took the position that they were only reloading discarded camera shells. Therefore, the central issue before the ITC was that of "repair versus reconstruction." Were the respondents carrying out a manufacturing process extensive enough to constitute the making of new cameras, or were they merely repairing used cameras?

In the first two months of my new consultancy, I visited Stroock's offices twice, looked at half a dozen videos taken during factory visits in China, examined 15 to 20 infringing cameras representing the various camera types, and began the process of writing a report on my findings. During my first visit, I met several attorneys working on this case.

There were a total of 15 Fuji patents, three of which were design patents, one was a method patent, and the other eleven were device patents. The design patents only covered a specific appearance. The other 12 patents included a total of 29 independent claims, each with 8 to 15 elements. There were ten different camera types being infringed, some manufactured by Fuji, others manufactured by Kodak and Konica under license from Fuji. DeCarlo had already developed pages and pages of claim charts. Each chart, sometimes two pages, covered a single claim from a single patent, each row in the chart represented a single element of the claim, and each of the ten columns represented a single type of camera. Altogether, these charts included about three thousand instances of potential conflict, each of which needed to be checked off as a *yes* or *no* infringement. Finally, these infringed claims had to be linked to each of the 23 infringing companies.

They supplied me with one sample of each of the ten different camera types. For each type, I opened the camera—sometimes with great difficulty because of the ultrasonic sealing of parts—and examined it for each of the possible infringed claims including each element of each claim. With about 3,000 potential instances of infringement this was no quick process, and I soon learned to sequence the patents so that one observation didn't obscure a later-needed observation.

REPORTS – My first cycle of writing the report on my findings was met with some challenge, quick learning, and adjusting my style to fulfill the legal requirements. But DeCarlo was happy with my refinements and told me the report was well written. Another attorney once said that my expert reports were clearer and easier to read than any others she had ever read. She even said that sections of them "read like poetry"—an exaggeration I appreciated nevertheless.

The report was finished in September. But several of the respondents submitted expert reports against which I had to write rebuttals. My first rebuttal report had three parts: opining on the issue of repair vs. reconstruction; specific rebuttals on each of the five expert reports; and addressing the question of validity of the '649 method patent describing two methods of film assembly. For the validity question, Stroock supplied me with "file wrappers" for each patent—a copy of the complete history maintained by the patent examiner. Most of them were two or more inches thick. These revealed every reference and citation that Fuji supplied to the examiner, and were used to defend against the charge that Fuji had hidden facts from the examiner. Also, Stroock sent me a collection of about 60 cameras, at least one of which had been remade by each of the 23 respondents. These were cameras that had been found on store shelves by their investigator. My responsibility was to verify that they were consistent with the respective camera types already studied and summarized. It took several days to work through this group.[19]

In addition, a deposition was scheduled in October, and I went to NYC a day ahead for some training and preparation. There were two or three attorneys that took turns grilling me for 10 hours over two days. I believe that the deposition went well, although DeCarlo kept reminding me (as did the Polaroid lawyers years earlier) to keep my answers brief and don't volunteer anything—an admonition that I was never quite able to meet because of my instinct to be "helpful."

THE FIRST HEARING – The hearing at the International Trade Commission was scheduled for the first two weeks of November. DeCarlo asked me to come a day ahead for preparation. Ultimately, I spent most of the first week in a suite of rooms with computers, documents, and lots of cameras. Much of the time was spent with DeCarlo going over procedures without actually preparing for testimony.

An unexpected undertaking during that week began with my discovery that a recently-found infringing, but newly-built camera by Achiever was slightly different from the first one I had examined. But to see the infringement clearly, a section of the camera needed to be carved away. My tools were 400 miles away—I needed to be resourceful. After purchasing a utility knife and a cheap soldering iron in a L'Enfant Plaza drug store, I used the latter to melt a rough hole through the plastic of the desired shape and the former to clean up the edges. As we combed through other Achiever samples, we ultimately found eleven different versions of cameras, some of which avoided one or two of its many infringements. Even the Achiever attorney didn't know about all these differences and embarrassed himself when he asked me to open a sample camera while on the witness stand to prove that it did not contain an infringing feature, which, in fact, it did contain.

On the occasions I was in court, progress was often slow because there were both Japanese and Chinese witnesses for which translations in both directions were awaited. On Saturday (my seventh day in DC), I was first called to the witness stand at 2 pm and stayed until 6 pm when court was adjourned and continued until Monday. On Monday, I wasn't called until nearly 4 pm, and no dinner break was called. I remained on the witness stand until court was adjourned at 10:10 pm. On Tuesday, I returned to the witness stand at 8 am and remained until 11:30. Altogether, I testified for nearly 14 hours. The Administrative Law Judge was a crusty older chap who periodically questioned me directly from the bench. Years

---

19 Sometime later and after I had created an organized checklist, I determined that a complete inspection of just one single-use camera took nearly an hour.

later my "handler" attorney told me that the Judge seemed to like listening to me—"It's not often they get to hear the intricacies of real hardware in court"—and to trust what I said—"He quoted you many times in his decisions."

Months later I learned that the decision was almost entirely in favor of Fuji, meaning that importation of these cameras must cease.

After the ITC hearing, I didn't hear much from Stroock until nearly a year later when I examined some more cameras and produced another report. A couple of years later, there was another trial at the ITC—this time it was an "enforcement" hearing to petition that the ruling was not being enforced by the Customs Department. By this time, all of the original respondents except for Jazz Corporation and Achiever had stopped US sales of reloaded shells. In fact, Jazz continued to ignore all the decisions along the way and was the lone respondent in the last trial I took part in some ten years later.

CONVOLUTIONS – There were so many cycles to this consulting job, so many cameras to look at, so many reports to write, so many issues to rebut, so many depositions, and so many days in court that I no longer remember the details. At the beginning of my consulting, the case was about the remanufacture of spent shells; then it evolved to newly-made cameras with similar, infringing features; then it evolved to damages suits, only one of which proceeded to trial; and finally it was about bankruptcy. Most of the original 23 respondents quickly fell by the wayside; Jazz didn't care; and Achiever diligently redesigned their cameras to get around each of the infringements, especially their avoidance of the film-loading patent.

It must have cost Fuji far more money than it was really worth. During each of these trials, there were typically four active attorneys from Stroock, a few more on the sidelines, two paralegals, a secretary or two, an attorney licensed in the state of the trial, and one or two computer experts. In addition, hotel rooms were booked to use as office space, office equipment was leased, computers installed and linked within the courtroom, and movers hired to deliver the scores of file cases kept nearby. My charges were a mere flyspeck. For some perspective on the magnitude of single-use camera products, when Fuji introduced them, their fondest hope was for sales in the range of 100,000 units per year. Over a million sold the first year and at the peak of sales some 200,000,000 cameras were being sold worldwide, a big fraction by the infringers. The final bankruptcy judgment in June 2009 was for $11,736,376. Seems pretty small for ten years of chasing.

There was great variety in my assignments over the decade-long engagement with Stroock and Fuji. At first, my part was to explore, verify, and delineate infringements seen in supplied cameras and later to opine on the patent application details for purposes of validity. I viewed over forty manufacturing videos, and even flew to Shenzhen, China in September 2003 to see factory operations for the purpose of opining on repair vs. reconstruction. In the first years I examined hundreds of single-use cameras. On one occasion I was sent 83 shells and given only three days to do a full analysis. "Are you kidding me? Each unit takes nearly an hour." But I organized them into similar product groups and managed to get it done and to find at least one infringement in every sample.

One year during our one week vacation on Sanibel, a dozen cameras were FedExed to me, and much of my time was spent preparing a report needed by Friday. In early 2005, I visited every pawnshop in Naples looking at 1970s-vintage cameras to verify film-advancement operation in preparation for an ITC rebuttal. At one point, a Fuji engineer and I spent two days in Stroock's offices examining 1,300 cameras to certify that a previously-marked chart had been correctly marked. Throughout the decade, I learned more detail about word processing than ever expected and generated over 250 pages of reports that required the use of styles, sections, automatic multilevel numbering, and sometimes complex chart production (*e.g.* those charts for 83 shells vs. 29 potential infringements). I even had to study and learn how a commercial film-developing machine worked.

That last need arose when Achiever finally invented a process to circumvent the film-loading patent. But their process was very cumbersome, and Fuji suspected that although Achiever demonstrated it in a video, they probably did not use it for the bulk of their manufacture. I was asked if there were a way to determine if they *really did* use that process. I reasoned that the evidence would be found in the light-struck patterns at each end of the film, but these areas were ordinarily lost or exposed again in the typical film-developing machine. My solution was to cut the film in half in a darkroom, wind each half into a reusable film cartridge with the original "middle" extending from the cartridges and without tape holding the film to the spool. After development in the machine, the ends would have avoided additional light-fogging or cutting and reveal the patterns I was looking for. True enough, Achiever did not manufacture many cameras using this method—they were still infringing despite their protests.

• • •

During my October 2002 appearance, the trial was the first damages case to make it to the courtroom. The case was against Jazz and heard by the US District Court, District of New Jersey.

Whether it is the modern lawsuit in general or this case in particular, the courtroom was packed with documents and computer equipment. I counted more than 49 transfer cartons of exhibits and other documents. Five benches were taken up with notebooks and other reference materials. There were 19 monitors, including laptops, many of them connected to the "live notes" being generated by the court reporter's stenotype machine. A large projection screen stood opposite the jury box. The carpet was covered with cables held in place with duct tape.

• • •

My last involvement with this case was for a bankruptcy trial against Jazz and its founder and owner, Jack Benun in June 2009. During my last 30 minutes on the witness stand, I testified to some recently-purchased cameras that had at least one infringement each—a true sign that Jazz and Benun were still making no serious attempt to avoid infringements after eleven years of being hounded. By this time most of Fuji's patents had run out, but the trial was for the purpose of collecting the amounts due from Jazz and Benun as a result of earlier court orders that preceded their bankruptcy.

At the end of that trial, I had been consulting for Stroock and Fuji for almost eleven years and was told that this was undoubtedly the end. It was.

This consulting engagement was a lot of fun. The topic couldn't have been closer to my likes, skills, and experience, and the latter made me truly expert and practically unimpeachable—as was attempted several times. There were times when I was overwhelmed, but frankly the panic gave me a bit of a high, and the thrill of finishing on time and in a way that was well received by Stroock and Fuji was satisfying. It was certainly fortunate that my flexible schedule at Sylvania made it possible to meet some of the deadlines. I found myself quite comfortable and self-confident in depositions and on the witness stand—I knew my stuff!

    I miss the action.

• • •

Court Appearances:

November 1998, International Trade Commission, 14 hours of testimony.

February 2002, ITC Enforcement hearing, 21 hours of testimony.

October 2002, civil trial against Jazz, Newark, N.J., 0.4 hours of testimony.

December 2003, ITC Enforcement II hearing, 9.1 hours of testimony.

June 2005, ITC Remand hearing, 5.2 hours of testimony.

August 2006, bankruptcy trial against Jazz and Mr. Jack Benun, 3.5 hours.

April 2007, bankruptcy trial against Jazz and Benun, 4 hours of testimony.

June 2009, bankruptcy trial against Jazz and Benun, 0.5 hours of testimony.

DANGEROUS and STUPID DOINGS

Three of my college friends died within a year of my last seeing them as a result of risky activities. Two died in crashes of their sports cars, and one died while rock climbing in the Tetons. I have tried to be more intelligent about my habits and activities. However, I wasn't always able to be so smart.

The Cow Pasture – When we were about ten, Gene, James, and I engaged in two dangerous activities at the manmade lake in Ashcraft's cow pasture. Probably in late summer or fall, we were on the outside slope of the earthen dam standing in tall, dry grass. James pulled out some matches and suggested we start a fire and put it out by pissing on it. It worked the first time, but his second strike wasn't followed with any more piss, and our efforts to stomp it out were for naught. The fire grew to about ten feet in diameter, smoke filled our eyes, and we were seriously panicked. Finally, I gathered my wits, broke off a green bough from a nearby cedar tree, began beating the border out, and ordered them to do the same but working in the opposite direction. When we met 180° later, the blaze was doused. Close call—and a lesson learned.

Either before or after the above stupidity, we were at that same lake in deep winter. There was ice covering the lake, and we ventured onto it starting at the dam where the water was deepest and the ice thickest. Well, deep winter in Charlotte wasn't enough to develop very thick ice, and when we got to the middle we began to hear cracking sounds. But rather than retreat, we panicked and headed for the nearest shore where the ice was even thinner. Needless to say we broke through, dropped into the water to our waists, but eventually made it to shore, cold and wet. Close call—and another lesson learned.

The Bear – In high school, I climbed Mt. LeConte for a third time. In the late afternoon after getting to the lodge, I heard or saw that there was a mother bear with her two cubs behind the dining hall. I was into photography at the time, pulled out my camera, and approached the scene. The mother was moderately close to the lodge, the two cubs well away at the base of a tree. I snapped a picture that included all three, but it also included far more. I wanted a better shot, but had no telephoto or zoom lens. So, I cautiously moved closer...and closer. I was about 40 feet from the mother and maybe 60 feet from the cubs when the mother decided that I was too close and took chase. Although her strides had a lumbering look, she was moving fast. I turned and began running away as fast as I could, noticed that I had no refuge other than to circle around toward her and toward the lodge, and could also see that she was gaining on me. I was frightened and without any obvious options. I abruptly stopped, turned toward her, raised my arms in a quick gesture, and screamed at her. She stopped, turned, and lumbered back toward her cubs.

But, what if she hadn't? I had lost my momentum and my distance. She would have been upon me in a second or two. I would be wearing the scars of that fight to this day. It scares me to even think of it.

Drafting a Truck – A few years later, the summer after graduating from college, surely at a time when I was much smarter, I was motoring from Charlotte to Boston on my Zundapp motor scooter. The trip took two days as the scooter had a top speed of about 45 mph. It was a beautiful early-September day. I was entirely comfortable, but was anxious for a long drive to be done with. Somewhere along Route 301 through Maryland, a four-lane but not very busy highway, an ugly dump truck passed going a little faster than I, about 50 mph. As soon as he was a few feet ahead of me, I changed lanes, pulling in behind him to use his draft. Sure enough, I picked up about 5 mph and followed him at 50 for about 10 miles. I was only 15 or 20 feet behind him. The entire time I kept thinking about what if he stops suddenly, what if he straddles an object in the road that I run into, what if I don't see a pothole ahead, what if..., what if... But I kept on following.

Fortunately, my comeuppance was relatively benign. After about ten miles, my machine suddenly developed a high-pitched scream and the gap between us quickly opened up. At first, I was startled and bewildered by how fast that truck could pull away from me, but in a moment realized that it was my machine slowing down. I pulled over onto the shoulder where the scooter abruptly stopped. It wouldn't crank. I soon reasoned that the engine had seized, a condition in which overheating causes the aluminum piston to expand against the iron cylinder wall until it locks in place, unable to move any more.

After puzzling for 20 minutes or so about what to do on this lonely road, I tried again to crank it. This time it did turn over, and it did start, and I did complete the trip without another incident. Apparently, the "helpful" slipstream wasn't so helpful for the airflow needed by that air-cooled engine.

A lesson learned, but what if...

Living Outdoors – The above events were recognized to be dangerous at the time. But I stupidly took the risk, anyway. But I have also done dangerous things that were only recognized as such years later from the perspective of age, wisdom, and experience.

One collection of such dangerous undertakings occurred during the summer trip that I embarked upon immediately after grad school. Not that the trip, *per se*, was particularly dangerous, but some of my activities during that trip were distinctly on the edge of reason. No wonder my parents worried. I surely hope my children and grandchildren don't do this kind of thing.

Many nights I slept in unsupervised areas. Sometimes I just stopped by the roadside, found an unused parking lot, drove behind a warehouse, discovered a clearing at the end of a dirt path, or, my favorite, settled in a quarry or gravel pit. I was never hassled at a gravel site, and it was always easy to shape the ground for my then-bony body to lie upon.

The only occasion that actually frightened me when it happened was when I chose to sleep by the roadside in the western Texas ranch country. I found a small rest area, parked my car in it, ate dinner there, but climbed over the barbed wire fence and spread out my sleeping bag on the grazing grass. (After all, the grass is greener on the other side.) I gazed at the stars as usual until falling asleep, but when waking at dawn, I was looking up into the eyes of a cow. I started, and so did she. (It may have been a bull, but the docile response would suggest otherwise, and I never took the time to find out.)

But I do sometimes think about the possibilities that troublemakers might have discovered me behind the warehouse, or cars may have driven over me in the parking lot or gravel pit, or that the clearing down a narrow driveway might have been some possessive person's front yard.

A variant on these after-the-fact alarms stems from my choice of sleeping directly on the ground. If weather was threatening or insects were bad, I set up the jungle hammock outfitted with a roof and mosquito netting. Otherwise, I slept in my sleeping bag directly on the ground (except for a thin sheet of polyethylene to keep the bag clean and dry). Only once did I think about the possibility that a snake might have joined me in the bag, and that was one frosty morning near Glacier point on the south rim of Yosemite Park. Yet again, this was not a campsite, just a large flat rock that I had discovered. It was uncomfortable both because of the rock and because of the cold. In the morning, the bag and surrounding ground were covered with frost. Somehow the coincidence of a frosty locale and the flat rock, on the likes of which I had watched snakes sunning themselves, made me think of the possibility.

But in retrospect, I have thought many times of the odds of rodents or other critters crawling into or burrowing into the stuffing to seek the warmth. As I recall, it never happened. But what if...

Others – There have been plenty of other unwise activities that could have caused injury or death that fortunately did not but were of less magnitude and, therefore, less memorable. There was the water tower in Warehouse Point, Connecticut next to our apartment that I partially climbed in the summer of 1962. I was cautious, but not smart. (Two years later, I met the man who built that tower. He was the grandfather of the bride, and the occasion was the wedding where I was best man to a fraternity brother.)

I suppose all those trips into coal mines in the late 1970s were not particularly safe—especially the 75 year old US Steel metallurgical-coal mine where we went to the deepest part of the mine to watch a mining machine robbing a pillar. *Robbing* gleans the last of the coal just moments before the roof and rock above come crashing down. I didn't fully realize where we were or what I was observing until I noticed that the normally 10-foot high roof was butting my hard hat and that every wooden support post nearby was broken in the middle or otherwise squashed to a much shorter shape than originally. "Gee... By design this area of the mountain is intended to collapse within the hour! What am I doing here *now*?"

I can remember numerous times while hiking in the mountains when I dared to venture close to a precipitous edge, or to leap boulder to boulder at a site where a slip could be dangerous or fatal. At Looking Glass Mountain in North Carolina, Bill and I walked toward the edge of one of those enormous faces that can be seen from the valley to curve ever downward until they become vertical. We were just walking with no ropes tethered to anchors above. We didn't stop until our shoes began to feel *about* to slip. A slip would have dropped the slipee several hundred feet.

While climbing Angel's Landing alone in Zion National Park during my first visit in July 1964, I was distinctly aware that it was only my grip on a rusty pipe or chain that stopped me from pitching back and falling to the valley floor. What if a sudden fright such as a bee sting or a bird swatting against my face distracted me into letting go? And a week later in Yosemite Park, I bushwhacked off the trail and ventured stupidly close to abrupt edges at the top of lower Yosemite Falls whence it crashes to the rocks more than 300 feet below. (In 2011, three hikers climbed over a guardrail at the top of nearby Vernal Falls and were swept to their deaths over the edge of that 317 foot drop-off.)

INVENTIONS THAT DIDN'T ...

What exactly is an invention? If you have a great idea, is that an invention? Or do you need to put it into practice? Or somewhere in between? Centuries ago, the US Patent Office required a model of an invention before they would grant a patent. Today, a well-written description that seems not to have been anticipated in whole or in part is enough to obtain a patent.

As you can see in *Appendix A*, I have a number of patents representing inventions that have formally been recognized as such. But you can also guess from the project descriptions over my life that I undoubtedly had a number of clever ideas that were never patented though some of them might well have been inventions. But what intrigues me enough to describe them below are those ideas I had over the years, did nothing with, but later discovered that they had been put into practice by someone else. You might ask, "Well, why didn't *you* get rich on those?" Well, some of them probably didn't make anyone rich. Some of them were so far afield of my talents that I couldn't have known how to start implementing them. And some were ideas that I might have developed, but didn't for one reason or another including the time commitment away from my day job.

Paving Groove - When I was in my late teens, probably 1958, my father hired a contractor to pave the driveway. Only the single-lane strip from the street to the turnaround area in back was to be paved. The turnaround itself was to remain gravel. I was at home the day it was paved, probably by plan.

I remember wondering ahead of time how the contractor would form the transition from the paved section to the gravel section. Specifically, I was curious about how robustness of the pavement edge would be accomplished. I had my ideas, but how would the pro do it?

After the crew arrived and began grading and preparation, I had a chance to ask the foreman about his technique for the transition. He said that they would "just feather it out." Since I had already thought about the matter, I suggested to him that they should dig a shallow, V-shaped trench across the drive so that the end of the paved section could curve down below grade level while maintaining a full thickness of strength. That way, the pavement should not fracture and break off with use.

His response was, "Nah, we just feather it out."

Well, it was not my driveway, and it wasn't my place to tell a professional how to do his job, so I dropped the matter.

Needless to say, the feathered edge soon broke off, the gravel eroded away from beneath the edge, and more pavement cracked and broke. We made a practice of periodically raking gravel into a little hill against that edge. It continued to slowly break away to a very ragged edge for the next thirty years.

Some quarter-century later, I was driving through a road-paving project when I noticed that at the end, a groove had been jackhammered across the road. My memory was jogged back to 1958. I thought of that hypothetical V-groove, I compared it with this rectangular groove (about two inches deep by two or three feet wide), made a note of its location, and visited the site again the next day to discover that, indeed, the groove was the transition from old to new without creating a feathered region, which so often cracked and chipped away.

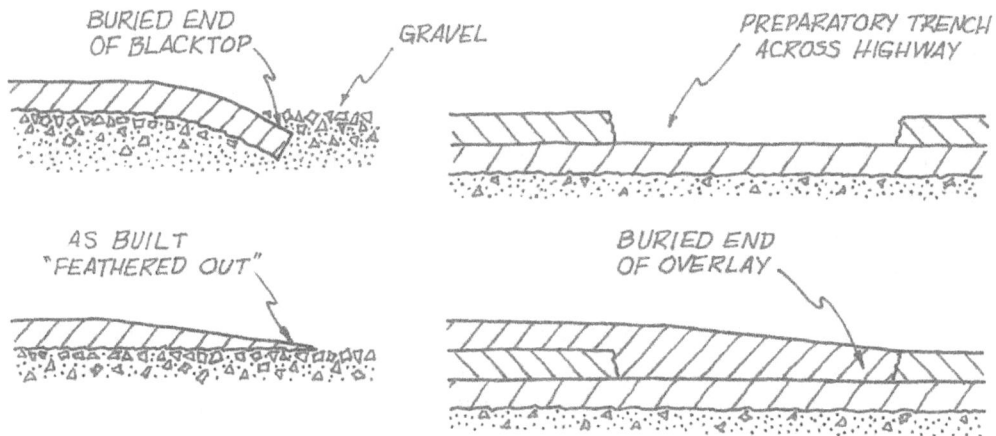

My vision of a proper termination, above. Its implementation, below—"just feathered out," which soon began to break away.

Preparation for a highway termination, above. The termination as completed, below, which is a near-simulation of the vision at upper left.

My "invention" was finally implemented. I have seen that method used many times since then. I doubt that it was patented or that anyone made any money on it, but obviously it was a good idea, and I had it first!

Inline Skates – As mentioned elsewhere, I discovered ice skating in high school. A few years later while walking to class at MIT across that long, cold Harvard Bridge every day, I began to think about having roller skates for the trip. Two things bothered me about conventional skates. Firstly, I had never particularly enjoyed roller skating after learning to ice skate because the foot had to remain flat to the ground, and secondly, once I got to school I would have to remove them to climb stairs and walk the halls. But I mentally worked on this challenge for months thereafter and developed a plan that used a comfortable boot, a rigid plate with a single row of four or five rubber-treaded rollers with good ball bearings, and said plate being hinged to the outside of the boot such that it could be latched in place against the sole or swung 180° to the side out of the way for ordinary walking.

Too bad I mucked up the "invention" with that hinging contraption and didn't just try to implement the rolling feature of it. I probably could have made a crude but usable pair in the student shop. And possibly be rich today!

Index Fund – While in graduate school, I learned that the stock market on average performed better than most portfolios or mutual funds. Upon pondering this, it occurred to me to start a mutual fund that did just that: buy a distribution of stocks that represent the market or some subset of it in weighted proportions and sell shares of that fund. Good idea? Well sure it was—but it was probably a good idea in the minds of a lot of other people as well. But how could an engineering student ever acquire the millions that even the simplest of startups would require? (At that time, it would have taken about ten million bucks just to buy a single lot of every stock.)

Some twenty years later I learned that Vanguard had launched just such a fund in 1976, using the S&P 500 as the subset of the market. I am sure they have made many, many gazillions on that idea.

Zipper Lane – While I was at Polaroid, the Southeast Expressway underwent an expansion program to add a lane wherever they could. The Expressway was a disaster that slowed to a crawl every morning and evening, and I thought that adding a single lane on each side wasn't going to help much. Since there was little or no room to expand outward, the gain was made by paving over the median strip and installing Jersey barriers. Upon reflection, it occurred to me that if the barriers could move, the roadway could be alternately widened in the direction of commuter travel twice a day.

My first thought was to install hydraulic pistons in grooves in the road and have the pistons push the barriers back and forth. I quickly dismissed this as impractical to say nothing of the annoyance of driving over all those grooves every few feet.

Next thought: design a truck to drive down the road, pick up a barrier, move it across one lane, and put it down. Too bad I never went the next step to refine that idea into something workable. I shared my idea with some of my work colleagues including my boss Dick Wareham. Ten or fifteen years later, Wareham sent me a clipping from a California newspaper that described just such a truck. The movable barriers had T-shaped flanges at the top and this real truck had the refinements of moving the barriers not one at a time, but having a continuous track of rubber tires rolling under the flanges to lift the barriers and snake them from one side of the truck to the other. Later I saw for myself the same system used at a construction project on the Connecticut Turnpike, and even later saw them installed on the Southeast Expressway, for which I originally had the idea.

Somebody undoubtedly *did* make a bunch of money on that machinery. Undoubtedly patented it, too. Too bad it wasn't me!

Camera-Phone Hook – Many mid-century cameras had a hook-eye or -eyes for a wrist or neck strap. Land didn't want to put hook-eyes on the SX-70 because it would make it difficult to slide into a pocket or purse. But while working on one of so many versions of cameras we designed, I noticed there was a bit of free volume in one corner. I had an idea. I devised a groove parallel to the plane of the closed camera that cut across one corner at 45°. The groove was about $3/16$ in. wide and about $3/8$ deep. Next a hole was drilled near that corner perpendicular to and fully through the groove, then a short dowel was driven into the hole and past the groove. Now there were no protrusions on the camera to snag in a pocket, but a lanyard hook could be snatched onto that pin recessed within the groove. A customer could have his cake and eat it too. If he were the pocket-type, then no wrist-strap and no protruding hook-eyes. If she were the swinger-type, then clip on the wrist-strap.

I thought it a neat and perhaps novel idea and had a simple model constructed. Undoubtedly an *invention*. But it didn't go anywhere and was never incorporated into a camera. But a generation later when cellular phones were still on the bulky side, I saw a few cell phones that had exactly my invention on it. I was very pleased that someone thought it a good enough idea to bring it to market.

Ice Barrier – When we had the Wayland house roofed in about 1980, I specified that the first course of roofing was to be a 3-foot wide roll of mineral roofing of the same color as the shingles. Then the shingles were to be laid over that strip to completely cover it. I was hopeful that it would provide a virtually waterproof barrier for the lowest 36 inches. I *insisted* that there be *no joints*. "Use a new roll at each edge—you can keep the scrap, but make *no splices*." Of course I recognized that it would have many nails through it, but that is the magic of thick, tar-based mineral paper—it typically doesn't leak around a tight nail.

Some years later I began hearing about a new product: an Ice Barrier, Ice Shield, or Ice and Water Shield. It was a three-foot wide piece of waterproofing material that was to be applied to the edge of roofs and under the shingles for just that purpose. Except for the formulation details, it served the same function as mine. I invented it first. But did mine actually work, or were we just lucky not to have had any significant ice dams in the next dozen years?

Well: About 15 years later we had a very hard winter and the house developed huge ice dams at every eave. And, wow—the kids' bathroom had such a big leak over the window that the gap between the window and storm window was virtually *filled* with ice. What a mess. And, what a disappointment that my plan didn't work. But, hey—there weren't any other leaks in the house. What did that mean? When spring came, I climbed up to that spot, lifted the shingles enough to inspect the first course of roll roofing, and found that the roofer had violated my specification—a joint. Exactly in line with the window. No wonder it leaked. And my "ice barrier" had worked, and I had invented it first.

"Cell" Phone – No, I don't claim to have invented the cell phone. When Motorola first announced this new technology it was called *cellular* service and the phones were *cellular* phones. The word "cellular" was hard for me to pronounce without getting my tongue twisted up so much that the word came out wrong. Soon I began calling them *cell* phones. At the time, I was working with the cellular phone people at GTE MobilCom and used the expression *cell phone* when talking with them. About six months later, I began to hear people away from the Labs using the same expression. I'll never know if I was the first to coin the expression or not, but perhaps I did start the trend through MobilCom, an organization that really could have moved the rolling ball along.

(Even if novel, an *expression* could never be an invention, *per se*. It might be used as a trademark or service mark, and thereby be protected by the USPTO, but not with a patent.)

Paving More Efficiently – When on the Road Construction Committee, I learned a lot about paving. I also learned how much repaving costs. When encountering road-paving crews in cities or on highways, it was always painful how much they tied up traffic. Putting all this together, I concluded there was a better way, at least in some instances. Less costly, too. Here was my plan.

Assemble a chain of machines as described below. They could be cleverly linked to follow precisely, be separately steered by a driver on each, or be steered by GPS to follow the identical track as the lead machine—any variant would satisfy. The first group of machines would be a series of five, ten, or maybe fifteen hovering ovens to heat the pavement until it was soft enough to be scraped up with a scoop. (I first saw such an oven on Boston streets in the early 1960s, but that one just worked a short section back and forth for many minutes until the pavement was soft.) The next machine would be a scraper to lift up the upper inch or so of the softened bituminous concrete (blacktop, asphalt). Next would be a gap with an overhead conveyor to carry all that lifted blacktop to a mixing machine. Within the gap one or more workers would be stationed to scrape up the loose material in the vicinity of structures (manholes, gate covers, catch basins, *etc.*). The mixing machine would mix the soft and hot but aged asphalt with additional liquid bitumen to bring it to the proper, *new-asphalt* consistency. The last machine would be a conventional paving machine that put the same material back in place.

But, to my thinking, there is an additional option, especially for city use. The chain of machines should be configured to process only half a lane at each pass. For one thing, the frequent structures typically found on city streets would most likely require that hand work be done to lift the soft asphalt adjacent to them. Hand work requires access, access suggests narrowness. Moreover, a six-foot wide chain of machines with a few surrounding workers would block up only one lane, not two. There would be no need for pavement-carrying dump trucks because the paving material comes directly from the roadway. The only supply trucks would be those for the liquid asphalt and fuel for all machines—delivery systems that would be more flexible, less frequent, and less intrusive. The extra pavement joint would typically be midlane between wheel tracks, a locus of low stress. Moreover, the preheating of adjacent asphalt along all seams would assure better quality joints everywhere.

Although this plan has been in my head for two decades, I always guessed it to be harebrained. However, in June 2012, I was driving through northeastern Kansas and witnessed a similar chain of events. The system covered a full lane; the heaters were on 8 or 10 separately-driven trucks; there were some intermediate milling operations; and there was a machine taking the hot pulverized pavement, adding a "rejuvenation agent," and passing it along to a conventional paving machine to be spread smoothly back in place. A typical roller compacted it. The rural highway didn't have any structures in the lanes to work around. I subsequently learned that the contractor and inventor is Dustrol Inc., in business about twenty years, with core operations in six states from Texas to Nebraska, and Oklahoma to Colorado.

My invention has been implemented—at least partially. (Maybe, considering that I first envisioned this system 21 years earlier while serving on the RCC, it is, in fact, my invention that was telepathically transmitted to the founder of Dustrol!)

One more enhancement of my "invented" system: Since the entire process would be completed in a single pass, there is the potential of having a lane-marker detection system at the head of the chain that is coupled to a GPS record such that at the tail of the chain, a light sprinkle of white stones could be dropped onto the doctored pavement in a pattern that mimicked the original lines. The roller would press them into the surface, and the faint marking would serve as a guide for the line-painting crew to repaint the highway markers without the expense of surveying, measuring, and marking from scratch each time the roadway is paved.

What's Next? – To be sure; I've had many more ideas than just those eight above. Those are the ones I can cite because they occurred twice: once in my head, once in the real world. I cannot predict which of my other crazy ideas might actually come into being—I look forward to seeing another.

• • •

So, why didn't I implement any of these inventions myself? Basically, I was content to work for an employer who appreciated my talent, paid me well for it, and supplied the money, resources, and support staff to accomplish our objectives. Over the years I carried out many inventions for those employers or consulting clients, 33 of which are memorialized with patents, many others of which simply became part of the product or project that I *did* implement. Others are incorporated into the projects around our homes or in houses that I remodeled. I am not the least ashamed of not having *also* implemented a zipper lane, or a paving system, or any of the many other ideas not even listed here. I performed my craft well and am proud of it.

RETIREMENT

Unless you have skipped ahead to this chapter, you must realize how much I loved being an engineer. I spent most of that life expecting to continue working until I dropped, *i.e.* I wasn't interested in or looking forward to retirement. But then a totally unexpected feeling came over me soon after turning 64. I was tired of my job and felt ready to quit. Would this have happened if my daily work had been true engineering instead of trying to apply and sell unwanted LED modules? I will never know.

But were I to retire immediately, there would be significant benefits left on the table—a lifetime pension, a medical insurance group plan, and of course another year of salary. So, I hung in there. The job wasn't that onerous. And I did have the satisfaction of designing one more new product that year.

My last day of work was late January a few days after turning 65. My wife joined me at a retirement luncheon with about forty colleagues. We soon departed for our two months of winter in Florida.

Perhaps the sudden transition from a four-day job to retirement was made easier for me by going directly on vacation instead of just hanging around the house all week. Vacations are supposed to be away from the jobsite. It softened the blow, if there ever was to be one.

So, what do I do with my time in retirement? To quote a fellow retiree, "I've gotten slower." It takes less stuff to keep me busy all day.

During those first two months of "vacation" in a rented condo in Naples, we had several visitors join us for periods up to a week; played golf half a dozen times; went often to the driving range in hopes of improving my lousy game; explored the area; discovered restaurants and other entertainment sites; met with other snowbirds from Massachusetts; experimented with several techniques for getting our email; began thinking of expanding the Cape house; did our taxes; read more books than usual; and I continued to consult. At the time, the Fuji infringement case through Stroock was still periodically active and some cameras were sent to me for evaluation. When we returned to Massachusetts, there were the springtime projects around both of our houses to clear them of winter debris, encourage the reemergence of plants and lawns, and all the other general projects the likes of which had kept me busy evenings and weekends for decades.

In 2005 and '06, we spent a lot of mental effort preparing to move from Wayland to the Cape. I was convinced that the Cape house at half the size of the Wayland house could be made suitable for living full time with minor tweaks, but my wife was unyielding that more space and a swimming pool were essential. We outlined a number of options for the Cape house, and I actually cut-in a new door, reversed the basement stairs, and constructed attic stairs with the hope that easy storage above and convenient access to a future finished game room, office, third bath, laundry, and two optional sleeping areas in the walkout basement would make the difference. We developed specifications and sketches, and met with three builders about bumping a little here, a little there. That didn't look too good, so we designed an addition with master bedroom, bath, side entrance, laundry, and a replacement garage. I hired an engineering firm to survey the yard and design a tentative septic system for four bedrooms. We even hired an architect to evaluate our design and tune it up. All of these activities took hundreds of hours as I worked on designs; purchased and studied the building code; picked the brains of three building inspectors; drove to the Cape and back many times; found building, septic, excavation, and basement refinishing contractors; met with them multiple times; and built a $1/8$ inch to the foot scale model of the contoured lot, extant house, and several models of alternate additions, decks, and pools.

We consolidated our investments, and I developed better methods of recordkeeping. Since my late teens, I had kept records of bank accounts and investments, but the old handwritten methods were cumbersome, obsolescent, and probably obtuse. The new records, summaries, and vital data for my survivors are now in much better shape.

Speaking of financial records, perhaps it is worth reflecting on my philosophy described earlier about saving money, analyzing opportunities, and investing aggressively but wisely. Applying it over a lifetime of ordinary engineer's wages has given us—six decades later—a comfortable retirement for my family with little financial worry.

Once we decided to purchase the Osterville house in late 2006, there were many projects undertaken to sell both the Wayland and the first Cape houses. Hundreds of hours were spent disposing of unneeded items through yard sales, Internet, consignment shops, family, and the dump; making last-minute repairs; and getting the houses listed. Our only successful realtor for both was our son and his newly formed Mayflower Properties, LLC.

Even when hiring projects to be done, it has always been my practice to thoroughly specify and oversee them, and with more time available in retirement I seem to have increased that involvement. For example, with the pool project, as explained earlier, we studied, discussed, and tested many configurations against potential future projects such as a larger deck. Finally, we had a complete and detailed specification. I hung around for most of its installation to make sure that each step was properly carried out and that they not crush our septic system or cut our buried phone and video cables.

In addition to the projects already described in the *Our Home in Osterville* section, others that kept me busy over several years included setting up a shop area in the basement; acquiring and assembling an unbelievably heavy but wonderful new table saw; struggling with a succession of four dehumidifiers to keep the basement dry; developing and implementing a check-list procedure to completely shut down the house each winter; installing a drywell for one downspout; laying 250 feet of drainage line for most of the others (with some hired digging help); designing and installing a lift-platform for getting awkward items to the attic; designing and installing over forty storage shelves in the attic, garage, and basement; planning the kitchen project enough to install one wall of cabinets two years ahead of time; getting a plumber to isolate the outdoor water lines with a backflow preventer, install new pipes to the sillcocks, and add two new sillcocks; converting an unused Freeze Alarm to a basement floor water-alarm; replumbing master bath and kitchen sink for speedy hot water; clearing pathways through the woods, removing dead trees, and consolidating brambles at a hidden dump site; making a ladder-hanger system in the garage for my three extension ladders; converting the dining room to a den with a custom 8-ft desk, double-doors to the living room, and a pull-out sofa for extra sleeping capacity; making and installing a hidden clockworks for the sunroom; and maintaining the yard with its unending regimen of weeding, seeding, trimming of trees and bushes, and readjusting flower beds. I like to keep busy.

A photograph of the author taken in the fall of 2012 at age 72. (Photo by Patrice Gilmore.)

Soon after moving into our new condo in Florida in early 2008, I was volunteered to help implement the new gate system for the community. That soon led to joining the Board of Directors; analyzing and overseeing the installation of the geothermal pool heating system; and later becoming president of the association. Whether that *requires* a lot of my time or that I just *choose* to invest it is another matter, but it keeps me busy. The fact that in the decade we've lived in the condo association we've managed to occupy two different condo units has kept me doubly busy with all those improvements needed to make our space convenient and pleasant for both of us. For two years, I volunteered and swung a hammer at Habitat for Humanity until the project site was built-out. For two seasons, I was a volunteer cart driver at the ACE Classic Golf Tournament. I got interested in radio-controlled sailing, bought a Soling kit, assembled it with much gnashing of teeth, and race it weekly with a group of retired gents.

Throughout the dozen years thus far of my retirement, I have done more book reading than previously and watched more TV, but only in the evenings. I had a brief cycle of volunteer landscaping at a local park, and for several years repaired books for three local libraries. And we now have grandchildren that we like to visit and kid-sit, occasionally for a week at a time.

My consulting for Tattletale users tapered off a few years after retirement and that for Stroock & Stroock came to an abrupt end after their last trial in 2009. I miss the scurry of Stroock assignments and the challenges of programming, but not too much. In the winter of 2011, I was stimulated to write *Reflections*, the original name of this book and put hundreds of hours into it. Five years later, I undertook editing it into *An Engineer's Life*, a shorter and more public book. I will miss the writing of my engineering life and experiences.

And, once again, my pace has slowed. What I used to do over the weekend now takes five days longer!

PUTTING IT ALL TOGETHER
As seen from 2012 or '18

Surely, I must be one of the luckiest persons in the world. I was born in the United States. My parents were healthy, well educated, of upper middleclass background, and of a hardy gene pool. I was born male, without deformities, with above-average intelligence, and with brother Bill to obviate being an only-child. My last name is distinctive but neutral and uncomplicated. And, I was born at the dawn of some very interesting times.

I was blessed with caring parents who provided a stable home life and made sure we got a good education from public schools in an egalitarian setting. Since early childhood, I knew what I wanted to be when grown up—and that's what I became. And engineering has always been in demand. I was fortunate to attend a high quality university without being saddled with debt. Although college was a major challenge for me, I graduated on schedule with at least some *A* grades.

I grew up to be tall, but not too tall. I can eat anything in sight without gaining weight—not much, anyway. I've kept a full head of hair and still have all...well, most of my teeth. At 78, my hearing and vision are still good. I haven't had hay fever for decades and rarely have headaches, stomachaches, or get sick. Back and joint pain, though bad at times, is usually minimal. I managed to avoid drugs and tobacco, although alcohol has sometimes been troublesome. I take no prescriptions and only two regular supplements.

My children are intelligent, healthy, avoided the pitfalls of youth, and are leading successful lives. They all own their homes, and three have wonderful and caring spouses along with two beautiful and intelligent children.

Most of my jobs have been largely exciting, challenging, and rewarding. And although my salary was that of an engineer and some of my investments went south, I have managed to develop enough of a nest egg that we should be able to live out our retirement years comfortably.

So, what's the hitch? Was it all sunshine and roses? Well, no, it wasn't. Since my college days, I have periodically suffered extended bouts of depression that get in my way and slow me down despite my efforts at keeping up a good face. While in these funks, I view my accomplishments as being insignificant—film-based photography is obsolete; the Polaroid Land Camera is long gone; those products designed for ECA were simplistic; the laboratory instruments designed at GTE were short-lived; the premature effort of introducing LEDs at Sylvania was a dud; *etc.* And during those periods I feel profoundly sad about so much—a family picture makes me wonder if I have been a good parent; another picture from younger days makes me feel the losses of passing time; glimpsing a camera reminds me of all that photographic life and those many photos that will never again be enjoyed; and a trip to the attic or basement where I see all that stuff makes me think of all those well-intentioned projects that will never get done.

Fortunately, these morose and pessimistic feelings pass with time, and I return to feeling well again.

• • •

How do I characterize my life's work? When asked about my profession, I usually reply *I am an engineer*. If asked for more detail, I might add *mechanical* engineer, *electro*mechanical engineer, product *designer*, *inventor*, or *optical* engineer. When I was about ten and was returning yet another lamp, toaster, or other small appliance to Mother after repairing it, I told her that I wanted to become a "fix-it man" when I grew up. And I sometimes answer the above question that "I fix things." After all, at over a decade into retirement, I cannot accurately answer that I still do many of those "engineering" things, but I do still fix things. Just think of the things I've fixed over the past 65 years: appliances, furniture, wallpaper, cars, sink drains, bathrooms, kitchens, spilling gutters, roofs, decks, at least 19 heating systems, door locks, flooding basements, house additions, shop floors, siding, a vault door, heat without power, lived-in look with randomized lights, winter shutdowns, a neighborhood entry gate, *etc*. And professionally, I have "fixed" an inverted broom and a surgical instrument at MIT; numerous camera challenges at Polaroid; both stagnant and new projects at ECA; a flexible drill shaft and a publicity movie at FMA; four more cameras at Avant; three space flight payloads, an early version of FiOS, and battery functionality at GTE; ceramic molding and an integrating sphere at OSI; and legal issues for both GTE and Fuji.

• • •

Did I like my life as an engineer? Am I proud of my own professional accomplishments? Did I truly accomplish anything? What part of it did I like best? What would I have liked even more to have done? How do I feel about my life's work?

Yes, I **did like my life as an engineer**. Even before I knew the word *engineer*, I wanted to be one. I liked to plan, design, and build projects from a very early age. I liked watching my grandfather at work in his shop and helping Dad around our house; and I sought out chances to watch houses being built, wells drilled, bricks laid, architects at work, furnaces installed, plaster laid, *etc*. In college my most enjoyable courses were engineering and design courses—I did best in those, too. For the most part, I loved all those professional jobs, and when I came home, I undertook similar avocations. I wouldn't have written this book if I hadn't liked my life as an engineer.

And, yes, I **am proud of my accomplishments**. Looking back on that early project of developing the SX-70 camera makes me very proud. Some of my optical insights and the construction of the *Len's & Bellows' Camera* in just a few weeks still impress me today. I feel somewhat responsible for and guilty about the camera's nearly two-year false start, but someone much bigger than I was controlling that. The camera may be dead and gone today, but it was once widely enjoyed and required Polaroid to spend $1 billion to finish what I started, and they ultimately sold more than 100,000,000 cameras. At ECA, I quickly got several stuck projects unstuck, designed some new products, managed one big money-maker into production in record time, and tuned-up a factory's production efficiency by nearly 20% over the summer. I accomplished a lot at ECA despite a difficult and erratic boss. At FMA, I was able to get the four-year-old flexible drill project out of the lab and into two different coal mines where it drilled over a mile of holes. Made a movie of it too! And at GTE, there was the Space Shuttle payload that I conceptualized, designed, built, tested, and managed from nothing to delivery at Kennedy Space Center in ten months. When I took up programming microcomputers, I wrote successful code that accurately controlled a number of systems and experiments over a decade's worth of projects. When serving as expert witness for GTE and Fuji, my work contributed to winning cases, and although I was challenged many times and impeached at least twice, I always prevailed. And, when I have given service to my family, friends, colleagues, and the community it has usually been received with genuine appreciation.

Perhaps the **best moments of being an engineer** were the little tidbits of interesting or even intriguing experiences, mostly at the *outside* of my job. For example, what stands out most from my summer job at Duke Power? There are flashes of the camaraderie, the education about electrical power taught by practitioners instead of scientists, the satisfaction of performing responsibly, and developing the respect of my peers and supervisors. But the memory that most stands out was the day we went to the Great Falls plant located at a dam that I had admired since childhood. That alone was a moving experience, but the real thrill was climbing down a narrow, rusty ladder 30 feet into the bowels of a power system and sitting inside the runner of the water turbine that was being serviced.

And what stands out from my days in college? One might expect me to cite the first-rate education I got, or the brilliant professors I studied under, or the lifelong friendships that began at Sigma Chi. And, of course I am thankful for all of those. But the distinctly rewarding and most memorable experience was being treasurer of the fraternity. Just being asked to serve was a huge honor and sign of trust in my integrity by my brothers. I took the job seriously, applied myself to it, performed it to the best of my ability, and generally showed myself that I could do most anything to which I applied myself. It built great self-confidence in me.

During my summer job at Scott Paper the actual engineering work was not all that exciting, and while I was working with talented people, perhaps the most intense and memorable event happened on a day away from the office during a visit to the Chester plant. A web of paper suddenly broke, and I watched a well-trained and practiced group of craftsmen converge on the problem site, each knowing his precise job without needing direction, and the problem was efficiently resolved in moments. It was not only a thrill to watch, but it taught me something about the inherent value of running an operation with truly competent workers who are proud enough of themselves and their jobs to perform them to perfection.

And speaking of pride, I must recount a similarly memorable event I saw during my first visit to London. Amid all the excitement of seeing Westminster Abby and the Houses of Parliament and St. Paul's Cathedral and the Tower of London, an insignificant event occurred one nice June day when I was standing on the sidewalk waiting for someone. I was idly watching a man in his fifties with a push-broom and trash container as he swept the gutter along a 40-foot stretch at a street corner. I will never forget the way he looked back on his completed job, tossed his head proudly, and painted a subtle expression of satisfaction across his face. This man had probably been working this job for the City for decades, knew how to do it well, and despite his humble station was going to keep doing his work to perfection until his dying day.

And Polaroid? Where else would I have earned 17 patents in six years, collaborated with so many fascinating and capable people, worked for an incredible personality like Edwin Land, been sent to Ansel Adams' Workshop, invited to cocktails with a Nobel Prizewinner, designed the board room table for a *Fortune 500* corporation, and had access to their wonderful collection of photographs by great artists?

ECA handed me the opportunity of fast-tracking many projects and several products into production. It was a fulfilling experience to accomplish as much as I did during the four years as chief mechanical engineer. ECA also gave me my only experience with industrial espionage when I was sent to a factory to learn, on a friendly but unofficial basis, how they assembled part of a solid-state sensor. In my last six months working as manager of manufacture, I significantly increased the efficiency of the Puerto Rico factory by rearranging the schedule to consolidate similar products into a contiguous production sequence.

Foster-Miller was a mixed bag. Some of the engineering work was challenging and satisfying, especially in the last year, while the proposals, reports, and meetings for and with government clients quickly became a real drag. But the outstanding memories from that period were the trips to coal mines, seeing the vast differences among them, meeting provincial and earthy people, and tasting the thrill of danger in the underground air. During that field test the last summer at FMA, I didn't have a legitimate reason for staying all six weeks, but I sure looked for excuses just to linger, crawl in the muck and darkness, talk to the miners, ride the coal belt, visit the active face, and smell the coal. It reminded me somewhat of the basement smells I learned to love when only a child.

My GTE job was my longest. There are lots of memories, but what about the high points? Certainly the first space payload was a high point and my ability to manage a complex project, direct others, call on my own creativity to resolve many issues, and pack it all into ten months were not only satisfying to me but caught the attention of others. But maybe the best moments of that project comprise the collected memories of silkscreening the names of several dozen contributing colleagues on the bottom of the payload, flying the payload to Kennedy Space Center in the corporate jet, working amongst the Utah State college kids at Patrick Air Force Base while preparing our payload, watching the launch, giving slide-illustrated talks at Goddard, the Labs, and several Sylvania sites, and meeting so many interesting people along the way including a few astronauts. A final but subtle accolade from this project came many years later when I discovered that *my payload* was pictured on a US postage stamp.

The highlight of the second payload turned out to be learning to program microcomputers. Until that time, programming had been a rare and onerous undertaking for me with minimal utility of the result. But now, I could program the many functions needed to heat and accurately control the temperature of a crystal growth experiment. The ability quickly branched to programming other needs: the monitoring of other payload temperatures, performing a linear regression analysis in real time of the three-axis accelerometer data, controlling the furnace's preparatory bake-out process, and other functions. Later, that ability resulted in a host of projects for Telops, MobilCom, NASA, ABell Engineering clients, Sylvania, and our own houses in Wayland and Osterville. It also allowed me to conceive, design, and build the Back Watch monitor, that very fulfilling final achievement at GTE.

I found those many internships with young people very satisfying. For the three students sponsored by the United Negro College Fund over three summers, I planned ahead for their two-week projects, and two of them accomplished a lot and seemed grateful. The two advanced-degree graduates in the Associates rotation program were enthusiastic, and ended up being quite helpful to our projects. One was especially supportive, and I watched him become quite active and popular among the younger employees.

Most of those telephone field trips were enjoyable and fulfilling, especially when they resulted in actually resolving a problem such as the battery failures and the video dial tone power reconfiguration. Most of the people I met at Telops or in the field liked having help from the Labs.

And probably the most memorable event while at Osram Sylvania was the trip to the Jefferson Memorial to troubleshoot an LED installation that was in trouble. The memory was undoubtedly made so indelible by the fact that I saw the Pentagon from the dome of the Memorial and the Twin Towers from my seat in the returning plane only a few days before they were attacked on 9/11.

Looking back on it all, **what would I like to have been different?** That's a hard question because basically, I liked doing exactly what I did. I do, however, have three thoughts on that question.

First, I should have stuck with product design. Polaroid may have been the antithesis of ECA, but at both, I thoroughly enjoyed the rigor of designing products. Designing for production requires a different perspective than designing small-scale, one-of-a-kind projects. Also, designing for the outside world is very different from designing laboratory or factory equipment. At the time I joined GTE, our Lab was serving as a product design center for some of its divisions, and I was hopeful of getting back into product design.

Sadly that came to an early end and ever after I found myself designing mostly laboratory equipment of various kinds including those space payloads.

Secondly, I sometimes wish that I had done something physically big and more permanent that could be pointed to with sustained pride. I might see an unusual building and think: I wish I had done something like that. Or cross a spectacular bridge and wish that I had built it. Of course part of the answer is that those projects were accomplished by huge teams, not an individual. But the real answer is simply that I didn't want to. I didn't want to work with a huge team. I never looked for the limelight, was always happy sitting in the back of the room, and was ever content to give my subordinates much of the credit for a project well done. Although I was happy to enjoy alone the satisfaction of having finished a project well, I do occasionally wish that some of them had come with better markers and public fireworks to announce that, in fact, my perception of a well-finished project was, indeed, a well-finished project.

Thirdly, I sometimes wish I had become a manager—not that I ever wanted to push people around, but that as a manager I might have been able to accomplish more by having more hands to carry out my interests and ideas. At ECA, I became a manager, but it was an odd position without personnel responsibility, in the wrong field, and it didn't last long. At Foster-Miller I was labeled Program Manager, but that meant being the lead report-writer and presentation-giver without any real responsibility for staff, project direction, or goal setting. At GTE, the major focus was research, and my practical nature might not have been very effective as a manager of science. As a result, I never lobbied for a management role, and the one time my director and I discussed such an opportunity, I ultimately demurred. In time, I was promoted to Staff Engineer, a technical track position equivalent to Manager.

How do I feel about my life's work? Turn back the pages and read! I feel good about it. The work was the right profession for me. There were rarely times that I was idle, uninspired, or depressed by the work itself. Typically, I found myself working hard, multitasking, stimulated, and creative. And being under self-imposed time pressure was like having a "runner's-high." I was constantly busy, felt good at the end of the day, and enjoyed the camaraderie with my subordinates, colleagues, and supervisors.

And some of that high level of interest, engagement, and excitement was probably the result of my "jack of all trades" style of engineering versatility. My degree was mechanical engineering, but I also had great interest and experience in optics, electronics, materials, mathematics, programming, manufacturing, broad concepts, small details, drafting, writing reports, making presentations, drawing illustrations, organizing projects, *etc.* This style perhaps left me as "master of none," but it surely filled my professional life with variety, enjoyment, and success.

I am also glad that at the end of the day, the week, or the year I could come home to a happy family, an interesting adventure, or a pleasant vacation together.

Appendix A
PATENTS, PUBLICATIONS, and REPORTS

PATENTS

Unless otherwise noted, the sole inventor is Alfred H. Bellows. Patents are listed in order of filing date. All patents are assigned to the previously-cited assignee company.

3,418,907 "Photographic Camera Erecting System," (the self-erecting bellows!), Dec. 31, 1968, assigned to Polaroid Corporation, Cambridge, Mass.

3,468,229 "Photographic Camera including a Scanning Exposure System with Compensation for Cylindrical Perspective Distortion by Optical Path Length Changes," Sept. 23, 1969.

3,505,943 "Photographic Apparatus," (motor rotor inside spread roller), April 14, 1970.

3,468,230 "Photographic Camera Having a Scanning Exposure System with Distortion Compensation," (similar to '229), Sept 23, 1969.

3,683,770 "Folding Camera," (basic SX-70 patent), with Edwin H. Land, Aug. 15, 1972.

3,498,194 "Camera Exposure Apparatus," (the homebody electronic shutter), March 3, 1970.

3,618,501 "Photographic Exposure Control Apparatus," (pivoting shutter blades), Nov. 9, 1971.

3,545,352 "Exposure Control Apparatus for a Photographic Camera," (shutter blades), Dec. 8, 1970.

3,557,678 "Exposure Control Apparatus," (pivoting shutter blades with edge cocking), Jan. 26, 1971.

3,554,076 "Compact Viewfinder with Toric Mirrors," (projected frame VF), Jan. 12, 1971.

3,614,412 "Photoflash Lamp Assembly," (flash array with 10 nested flash bulbs), Oct. 19, 1971.

3,622,242 "Stereoscopic Rangefinder with Movable Reticles," with Edwin H. Land, Nov. 23, 1971.

3,680,946 "Compact Rangefinding Device," (stereo rangefinder), Aug. 1, 1972.

3,610,123 "Collapsible Camera with Collapsible Viewfinder," (early style SX-70), Oct. 5, 1971.

3,610,128 "Stereo Rangefinder," (stereo RF shown on ColorPack style camera), Oct. 5, 1971.

3,619,202 "Variable Frame Viewfinder for Photographic Camera," (movable lenticules), Nov. 9, 19/71.

3,643,565 "Folding Camera with Developing Means," (door-closing circuit in SX-70), Feb. 22, 1972.

3,825,913 "Fuel Burner Supervisory System," (water-cooled extended scanner for large boilers), with Arthur G. B. Metcalf, Philip Guiffrida, July 23, 1974, assigned to ECA, Cambridge, Mass.

4,201,270 "Roof Bolter," (bolting machine outfitted with flexible roof drill), with Ribich, Hug, May 6, 1980, assigned to Foster-Miller Associates, Waltham, Mass.

4,440,154 "Solar Energy Collecting Apparatus," (compliant support for absorber tube), Apr. 3, 1984, assigned to GTE Laboratories, Waltham, Mass.

4,464,641 "Circuit Breakers," (residential breaker with center-reference bimetal), Aug. 7, 1984.

4,472,696 "Circuit Breaker," (residential breaker with reduced sensitivity to wear and friction), with Chung, Sept. 18, 1984.

4,472,701 "Electrical Circuit Breaker," (high power breaker, improved trip), with Piejak, Sept. 18, 1984.

4,491,814 "Circuit Breaker," (high power breaker with trip improvement), Jan. 1, 1985.

4,953,938 "Optical Fiber Expanded Beam Connector," (glass molding with inherent centering), with Buhrer, Sept. 4, 1990.

4,838,041 "Expansion/Evaporation Cooling System for Microelectronic Devices," (cooling system utilizing liquid carbon dioxide), with Duchene, June 13, 1989.

4,955,686 "Optical Fiber Crossconnect Switch," (mechanism for reconnecting in/out array of optical fibers), with Buhrer, Carlsen, Cousins, Sept. 11, 1990.

5,173,678 "Formed-to-Shape Superconducting Coil," (structure for forming a coated and laminated coil), with Levinson, Dec. 22, 1992.

5,347,246 "Mounting Assembly for Dielectric Resonator Device," (passive support means for resonator), with Loughridge, Sept. 13, 1994, assigned to GTE Control Devices, Standish, Maine.

5,299,100 "Microwave Powered Vehicle Lamp," (light source for automobile head lamp), with Lapatovich, Mar. 29, 1994, assigned to GTE Products Corp., Danvers, Mass.

5,937,033 "Telephone System Diagnostic Measurement System including a Distant Terminal Drop Test Measurement Circuit," (system to simplify data collection for diagnosing remote telephone equipment), Aug. 10, 1999, assigned to GTE Laboratories.

5,920,802 "System and Method to Improve Power Distribution in a Coaxial Cable Amplifier," (simple jumper to allow power distribution from two separate sources), July 6, 1999.

7,052,649 B2 "Mercury Dispenser for Fluorescent Lamps and Method of Dispensing," (mercury capsule that bursts open after lamp is sealed), with Grossman, George, Keup, May 30, 2006, assigned to Osram Sylvania, Danvers, Mass. ("B2" indicates previously published.)

JOURNAL PUBLICATIONS

"Arc Discharge Convection Studies: A Space Shuttle Experiment" (with A.E. Feuersanger), *Proc. 1984 Get Away Special Experimenters Symposium,* C.R. Prouty, ed., NASA Goddard SFC, Greenbelt, Maryland, pp. 17-24 (NASA Conf. Pub. 2324) (1984).

"Convection and Additive Segregation in High-Pressure Lamp Arcs: Early Results from a Space Shuttle Experiment" (with A.E. Feuersanger, G.L. Rogoff, and H.L. Rothwell) *Gaseous Electronics Conf.* (1984); *Bulletin, American Physical Society 30,* p. 141 (1985).

"Convection and Additive Segregation in Metal-Halide Lamp Arcs: Results from a Space Shuttle Experiment" (with G.L. Rogoff, A.E. Feuersanger, and H.L. Rothwell), *Symposium on Science and Technology of High Temperature Light Sources,* Electrochem. Soc. Meet., Toronto, Canada (May 1985); *Ext. Abstr. 85-1,* Abstract No. 385, p. 551 (1985).

"HID Convection Studies: A Space Shuttle Experiment" (with A.E. Feuersanger, G.L. Rogoff, and H.L. Rothwell), *Lighting, Design, and Application 15-8,* pp. 30-34 (August 1985).

"The Comparative Study of the Influence of Convection on Gallium Arsenide Growth" (with J. Gustafson and J. Kafalas), *Proc. 2nd Pathways to Space Exploration Workshop,* Orlando, FL (June 1986).

"A Comparative Study of the Influence of Buoyancy-Driven Fluid on GaAs Crystal Growth" (with J.A. Kafalas), *6th European Symposium on Materials Science Under Microgravity,* Bordeaux, France (1986).

"A Payload for Investigating the Influence of Convection on GaAs Crystal Growth" (with G.A. Duchene), *Proc. 1987 Get Away Special Experimenters Symposium,* N. Barthelme, ed., NASA Goddard SFC, Greenbelt, MD, pp. 77–82 (NASA Conf. Pub. 2438) (1987).

"Systems and Applications Development for Integrated Evacuated CPC Collectors" (with J. O'Gallagher, R. Winston, and W. Schertz), *Proc. Annual Meeting of the American Solar Energy Society,* Cambridge, MA (June 1988).

"The Integrated CPC*: Recent Progress, Present Status, Future Directions" (with J. O'Gallagher, R. Winston, and W. Duff), *Proc. Annual Meeting of the American Solar Energy Society,* Denver, CO (June 1989). *CPC = Compound Parabolic Concentrator.

"Interface Demarcation in GaAs by Current Pulsing" (with D. Matthiesen, J. Kafalas, and G. Duchene), *Proc. 28th Aerospace Sciences Meeting,* Reno, NV (January 8-11, 1990).

"Free Float Acceleration Measurements aboard NASA's KC-135 Microgravity Research Aircraft" (with D. Matthiesen, and G. Duchene), *Proc. 28th Aerospace Sciences Meeting,* Reno, NV (January 8-11, 1990).

"Microgravity Acceleration Measurements aboard the Get Away Special Bridge during STS-40" (with D. Matthiesen), *Proc. 30th Aerospace Sciences Meeting,* Reno, NV (January 6-9, 1992).

"A Versatile Get Away Special Furnace for Materials Processing in Space" (with D. Matthiesen, G. Duchene, and G. Chen), *Proc. 30th Aerospace Sciences Meeting,* Reno, NV (January 6-9, 1992).

INTERNAL REPORTS while at Massachusetts Institute of Technology

"Servo Controlled Inverted Pendulum," by AB for the team, May 23, 1961, Course 2.671 Term Project. 12 typed pages plus introduction and illustrations.

"Servo-Controlled Inverted Pendulum," AB, Sept. 1962, Report No. M-8931-3, Engineering Projects Lab, Dept. of ME, MIT. 24 typed pages plus forward and 9 sheets of drawings up to 22x34 in.

"Design and Construction of an Optical Adjunct to Ultrasound Neurosurgical Equipment," AHB, June 1962. Submitted in Partial Fulfillment of the Requirements for the Degree of Bachelor of Science, MIT. 30 typed pages plus abstract and 6 sheets of drawings 17x22 and 22x34 in.

"Pedal Control for Automobile: A Device to Enable an Armless Person to Drive an Automobile," AB, June 25, 1962. Submitted to the Engineering Undergraduate Award Program for Arc Welded Designs of Machines. 13 pages total.

"Restricted Channel Creping: A Preliminary Study," AB, Sept. 6, 1963, Report No. 463, Engineering Research Division, Scott Paper Company.

"Development of a Tape to Tactile Braille Reading transducer," AHB, June 1964. Submitted in Partial Fulfillment of the Requirements for the Degree of Master of Science, MIT. 31 typed pages plus abstract, TOC, and 43 sheets of drawings, mostly page-size.

INTERNAL REPORTS while at Polaroid Corporation

"General Photographic Data for Engineering," by AB, January 1966. Includes Nomenclature, Lens-Image Equations, Magnification, Supplementary Lenses, Power of Lens, Aperture Sizes, Depth of Field, Depth of Focus, Standard apertures, Filters, Prisms, Rangefinder Base Length, Shutter Efficiency, Flash Synchronization, and a Chart of ASA film speed *vs.* Shutter Speed *vs.* Relative Aperture *vs.* Brightness (candles/sq ft). 8 pages.

"Exposure Nomograph." Designed for the scanning camera and includes five terms: in/sec, slit width, exposure time, EV number, and f-stop.

"Concept Reports." A collection of possible inventions by anyone on the SX-70 project, described and compiled by me, then submitted to the Patent Department. Includes about 56 disclosures of ideas from E.H. Land, Dick Wareham, Irving Erlichman, Phil Norris, Howie Rogers, Al Bellows, *et al.*

INTERNAL REPORTS while at Electronics Corporation of America

"Photographic Summary of the Extended Scanner." Nine photographs and 4 diagrams including a cutaway perspective illustration (tuned up by an illustrator following my original sketch). All done by AHB, Spring 1972.

"Duplex, Self-Check, Liquid-Cooled Extended Scanner," Engineering Report No. 15-92, by AHB, Sept. 1972. Narrative description of the scanner with specifications and diagrams. 18 pages.

INTERNAL REPORTS while at Foster-Miller Associates

"Pictorial Summary: Head/Tail Support," AB, December 1975, 14 pages, figures only of a dozen different concepts and variations, Project BM7522.

"Program Status Summary: Development of Remotely Operated Longwall Head/Tail Supports," Phase IA, AB, February 1976, 14 pages mostly illustrations, diagrams, and bulleted text.

"Supplement to Phase IA Report: Remotely Operated Longwall Head/Tail Supports," AB, April 1976, 14 pages.

"Design of a Retrofitable Temporary Face Support System: Final, Phase I Report," AB, John Curcio, December 13, 1976, 87 pages.

"Collet Data: Flexible Shaft Roof Drill: A collection of Miscellaneous Data and History for the Performance of a Stress Analysis," AB, January 1978, 7 pages including 2 figures.

"Photographs: Flexible Roof Drill Ready for Underground Test," AB, May 1978, 13 photos with captions.

"Design, Fabrication and Field Testing of Remotely Operated Longwall Head/Tail Supports: Final, Phase IB Report," AB, John Curcio, September 1978, 88 pages including figures and six A and B size drawings.

"Flexible Shaft Roof Drill." A five-minute narrated movie illustrating the project. Shot, edited, and produced by AB in the summer of 1978.

"Flexible Shaft Roof Drill: Report on Progress and Status," AB, December 1978, 49 pages, "A narrative report on the Flexible Shaft Roof Drill at the time of Bellows' resignation, outlining current status, immediate plans and actions, and thoughts for the future."

INTERNAL REPORTS while at GTE Laboratories

"Development of the Modular Circuit Breaker," by AHB, March 1982, 51 pages.

"Study of Convection-Free Metal Halide Lamps: Final Report on a Space Shuttle Experiment," AHB, Fred Feuersanger, Gerry Rogoff, Harold Rothwell, May 1985, 44 pages. (Get Away Special experiment aboard Challenger, STS-11 in February 1984, mostly written by me.)

"Development of a Space Shuttle Payload for study of Convection in Arc Discharges," AHB, March 1985, 14 pages plus 3 commendations and 11 public relations publications. (This R&D summary was submitted in support of my nomination for a Warner Technical Achievement Award, a prestigious award conferred by GTE Corporation, but not won.)

"Microelectronic Packaging and Materials Directions," Bowerman, AHB, Avella, July 1985, 11 pages (little written by me).

"Microelectronic Packaging Research: Final Report," Avella, AHB, Bowerman, May 1986, 23 pages (little written by me).

"Local Powering of the Optical Network Interface: Requirements, Issues, and Alternatives," AHB, Duchene, December 1990, 40 pages. (A powering study for fiber-to-the-home installations, written by me.)

"Battery Charger Testing System," AHB, Duchene, March 1992, 20 pages. (An operating manual for a testing system we made for GTE Mobile Communications, written by me. Revised in May 1992 to include drawings, parts list, schematic, code, and testing log.)

"A Comparative Study of the Influence of Convection on Gallium Arsenide Crystal Growth," AHB, Matthiesen, August 1991, 24 pages plus raw data. (Get Away Special experiment aboard Columbia, STS-40 in June 1991, mostly written by me.)

"A Comparative Study of the Influence of Convection on Gallium Arsenide Crystal Growth," AHB, Matthiesen, June 1992, 24 pages plus raw data. (Get Away Special experiment aboard Atlantis, STS-45 in March 1992, mostly written by me.)

"Prototype Optical Network Unit for Fiber-in-the-Loop Delivery of Telephone Service," Buhrer, AHB, Duchene, Wang, December 1992, 17 pages plus appendices of schematics and code. (A designed-from-scratch demonstration of an ONU, partially written by me.)

"Powering of ONUs with Bussed Cable vs. Individual Twisted Pairs," AHB, March 1993, 13 pages.

"Powering of ONUs in Small Groups with Bussed Cable," AHB, April 1993, 10 pages.

"Temperature and Humidity Measurements in DLC Cabinets: Phase I," AHB, November 1993, 6 pages. (A report of first 45 days of measurements from one of Florida's monitored sites.)

"Network Battery Maintenance," AHB, December 1993, 12 pages. (A summary of backup battery failures, causes, instruments available, and a proposal for custom instrumentation.)

"Bused Power at Branson: a Comparison," AHB, March 1994, 10 pages plus diagrams and results.

"Coaxially Bused Power at Branson: another Comparison," AHB, March 1994, 7 pages plus diagrams and results.

"A Survey of Pair Gain Systems for Fewer than 25 channels," AHB, Dakss, June 1994, 19 pages partially written by me.

"Temperature and Humidity Measurements in DLC Cabinets: Phase II," AHB, July 1994, 26 pages including graphs, which took more time than the writing. (A report expanding the Phase I report to cover the first 6 months of measurements from all nine sites in Florida, California, Wisconsin, and Michigan.)

"Temperature and Humidity Measurements in DLC Cabinets: Phase III," AHB, November 1994, 37 pages including graphs. (A report expanding the Phase II report to cover one full year.)

"Monitoring of Batteries in the Outside Plant," AHB, January 1995, 10 pages. (A preliminary description of the Back Watch monitor.)

"Temperature and Humidity Measurements in Underground Chambers," AHB, November 1995, 13 pages. (A study of temperatures at various depths within three 4-inch casings of different depths and buried on GTE property.)

"Battery Monitoring in the Outside Plant," AHB, Thompson, December 1995, 25 pages. (A progress report on the Back Watch monitoring system, written by me.)

"User's Manual: Back Watch Battery Monitoring System," AHB, Thompson, December 1995, 25 pages. (Manual includes details for installation, how to use, interpretation of data, operational details, and troubleshooting.)

"Back Watch battery monitor," A four-page handout summarizing functionality of the Back Watch, Revised December 2995 (prepared by me).

"Summary of 1995 Network Quality and Provisioning Support Activities," Wei, AHB, Dugger, Hefter, Moore, December 1995, 9 pages. (A summary mostly written by others.)

"Temperature-Buffered battery Vaults: A Review of Test Results from GTE Installations," AHB, Beaird, Sanders, September 1996, 6 pages. (A study of commercially available underground battery vaults.)

"Backup Power in the Outside Plant: Assessment of Two Alternatives," AHB, December 1996, 4 pages. (A review of lithium polymer batteries and flywheels for energy storage.)

"Battery Monitoring in the Outside Plant: Final Report," AHB, Thompson, December 1996, 42 pages including graphs. (Results from the Back Watch monitor, written by me, data logged by Thompson.)

"Expert Report for the Defendant, GTE *vs.* the local power company," AHB, 1995 or '96, a now lost-track-of report on behalf of GTE in York, Pennsylvania.

INTERNAL REPORT while at Osram Sylvania

"Maintenance of the Ten-Foot Integrating Sphere in Manchester, and a Description of its Repairs Completed in July 1998," by AHB, Aug. 15, 1998. 14 pages. ("A 10-foot diameter integrating sphere was failing due to mechanism failure, distortion, and inability to fully close. Its usefulness was restored during factory shutdown through replacement, repair, and tune-up of its casters, main hinge, shape-holding braces and pneumatic actuator. Long-term care and adjustment is the major focus of this memorandum. Repair parts are itemized and an exposition of the repairs and observations is also included.")

EXPERT REPORTS prepared for Stroock & Stroock & Lavan on behalf of Fuji Photo Film

NOTE: Stroock appended numerous exhibits to most of my reports, primarily diagrams and charts, but the page quantities noted indicate the narrative portion written by me. In a few cases, my work product also included tables. In addition to the listed reports, there were over a dozen declarations and witness statements generally prepared by Stroock but using my text.

Expert Report for the Complainant: Fuji Photo Film Company, Ltd., by Alfred H. Bellows, Mechanical Engineer; ABell Engineering, Wayland, MA 01778, at the US International Trade Commission, "In the Matter of Certain Lens-Fitted Film Packages," Investigation No. 337-TA-406 September 17, 1998 – 37 pages.

Rebuttal for the Complainant: at the ITC, 337-TA-406, October 2, 1998 – 12 pages.

Expert Report for the Plaintiff: in the US District Court, District of New Jersey, vs. Jazz Photo Corp., Jazz Photo (Hong Kong) Ltd., Jack Benun, Defendants, Civil Action No. 99-2937(FSH), March 26, 2000 – 27 pages.

Expert Report for the Plaintiff: in the US District Court, District of New York, vs. Charles Randolph Co., The Complete Wedding Lab, and Custom camera Design, Inc. Defendants, Civil Action No. 99-CIV-4535 (SHS), October 2, 2000 – 23 pages.

Expert Report for the Complainant: Consolidated Enforcement and Advisory Opinion Proceedings at the ITC, September 6, 2001 – 28 pages.

Rebuttal for the Complainant: Consolidated Enforcement and Advisory Opinion Proceedings at the ITC, November 7, 2001 – 8 pages.

Supplementary Expert Report for the Complainant: Consolidated Enforcement and Advisory Opinion Proceedings at the ITC, December 5, 2001 – 13 pages.

Second Supplementary Expert Report for the Complainant: Consolidated Enforcement and Advisory Opinion Proceedings at the ITC, January 25, 2002 – 14 pages including a drawing.

Expert Report for the Complainant: Enforcement Proceedings (II) at the ITC, October 1, 2003 – 52 pages including observations during factory visits and two charts produced by me.

Expert Report for the Complainant: Advisory Proceedings (II) at the ITC, October 23, 2003 – 11 pages.

Supplementary Expert Report for the Complainant: Enforcement Proceedings (II) at the ITC, October 29, 2003 – 11 pages including two charts summarizing each of 83 camera shells.

Second Supplementary Expert Report for the Complainant: Enforcement Proceedings (II) at the ITC, November 14, 2003 – 16 pages.

Expert Report for the Plaintiff: in the US District Court, District of New Jersey, vs. Jack C. Benun, Ribi Tech Products, LLC, Polytech Enterprise, Ltd., Polytech (Shenzhen) Camera Co., Ltd. Defendants, Case No. 05-1863 (KSH) (PS), June 1, 2006 – 9 pages.

PHOTOGRAPH APPEARANCES

Postscript 1955, Yearbook for Charlotte Country Day School. According to initials, 39 of the photographs were taken by AB.

The Charlotte Observer. Several sports pictures taken by AB at Charlotte Country Day School games were printed in the sports section during the 1954-55 season.

Evaluation Report on Work in Progress on Sensory Aids and Prosthetics, April 1964, Report No. 9211-2, Dept. of Mechanical Engineering, MIT. Most of the 20 photos were taken by AB.

Polaroid Land Photography, 1966 Edition produced by the editors of Popular Photography magazine. The photograph of bunchberries on p 4, above the Introduction was made by AB using Polaroid 55 P/N film and subsequently printed from the negative.

Flexible Shaft Roof Drill. A five-minute sound movie shot, edited, narration written, and titled by AHB, Summer 1978.

Many now-forgotten reports of mine and colleagues' extending from MIT through GTE included photographs taken by AB. Also many 35-mm color slides were made by AB for presentations by me and others while at GTE.

PRODUCT APPEARANCES

Time Magazine, June 26, 1972. The cover is a photograph of Edwin Land using and partially hidden by an SX-70, the camera originally designed by Land and AB.

Life Magazine, October 27, 1972. The cover is headlined "A Genius and his Magic Camera" with a caption, "Dr. Edwin Land of Polaroid demonstrates his new invention." The cover picture shows Land having just taken a close-up picture of two children that are reaching for the picture as it is ejected from the SX-70 camera.

Control Engineering, May 1982, Technical Publishing – Dun & Bradstreet Corp. The cover shows the Intelligent Industrial Relay (I^2R) designed by Malcolm McDonald and AB of GTE Products Corp. and which is a rack-mountable 12-I/O programmable controller (PC).

US Postage Stamp, 1996. The $3 stamp pictures STS-11 (incorrectly labeled STS-7) with the lid of the GTE-built metal halide lamp experiment payload shown as the middle of 3 unevenly spaced dark dots along the starboard edge of the cargo bay. The payload was designed and the project managed by AB.

Appendix B
RESUME and CURRICULUM VITAE

Probably my first resume was produced for those job-recruitment visits in little booths at MIT. Those early resumes were largely limited to courses, activities, interests, and theses. They were painstakingly typed and copied on the recently-introduced Xerox 914 copy machine, one of which MIT had available for a fee. None of my early resumes survive.

The resume partially reproduced below had its beginnings at GTE and was first generated by the Tech Pubs group using the then-new word processing technology. It was used in proposals for funded projects to which I was expecting to contribute. At some point I requested an electronic copy, which I kept up to date for some years. It was used when applying to Osram Sylvania and when I began consulting for Stroock.

This reproduction of my resume includes page 1 only—the other two pages listed patents and publications respectively that are also listed in *Appendix A*.

When consulting at Stroock & Stroock & Lavan, they needed a different style to include with their submission to the court regarding my credentials as an expert. That is when I first created a Curriculum Vitae. That is also reproduced below in its most recent form. Clearly it was focused on the issues of my technical and courtroom experience.

ALFRED H. BELLOWS
Wayland, MA 01778
508 555-5049

EDUCATION M.S. (Mechanical Engineering), MIT
B.S. (Mechanical Engineering), MIT

STRENGTHS Program management, project management, optoelectromechanical system design, design of mechanical structures and mechanisms, thermal finite element modeling, design of Shuttle-qualified hardware, conversant with Macintosh and DOS-based computers, Word, Excel, Canvas, Cadkey, and microcomputer programming.

EXPERIENCE **OSRAM SYLVANIA INCORPORATED (1997 to present)**
Staff Engineer: Mechanical design of new products including arc lamps, microwave powered lamps, and ballasts. Design of new and revised in-plant equipment for testing lighting components and products. Design and application of LED lighting modules.

GTE LABORATORIES INCORPORATED (1978 - 1997)
Staff Engineer: Program manager for electrical power systems projects including development of a remotely deployed microcomputer-controlled power monitoring system. Managed development of power-efficient module for remote interfacing of optical fibers to telephone equipment. "Payload Manager" of 3 space shuttle experiments, one to measure performance of a gravity-sensitive metal halide arc lamp in the absence of gravity, and 2 flights of a low-convection gallium arsenide crystal growth experiment. Developed advanced microelectronic packages including a high-performance heat-extraction module. Principal Investigator and/or contributor to several electromechanical development programs including a modular circuit breaker, a solid state programmable relay, a microprocessor-based industrial controller, an evacuated tube solar collector, and a high-current surge arrestor.

FOSTER-MILLER ASSOCIATES (1975 - 1978)
Program Manager: Management of and design responsibility for several US Bureau of Mines contracts to develop underground coal mining equipment. Activities included design of structures, vehicles, mechanisms, and hydraulic systems. Completed a flexible shaft drilling machine and tested it at two coal mines, where it exceeded the test goal of drilling 5,000 feet.

ELECTRONICS CORPORATION OF AMERICA (1971 - 1975)
1974-1975. Manager of Manufacture: Responsible for all production activities within Cambridge and Puerto Rico plants, and for supplying materials to Canadian and European subsidiaries. *1971-1974.* Chief Mechanical Engineer: Designed and/or approved all mechanical aspects of new commercial photoswitch and combustion control products. Project Engineer on three new products, carrying them from conceptual design into final production release.

POLAROID CORPORATION (1964 - 1971)
Senior Engineer: Designed various viewfinders, rangefinders, and shutters for ColorPack II camera. Designed experimental scanning camera prototype. Designed the basic layout of the SX-70 camera, its optical system, electromechanical shutter, spreading system, and several viewfinders and rangefinders for it.

MASSACHUSETTS INSTITUTE OF TECHNOLOGY (1962 - 1964)
Research Assistant: Developed first "E-Mail" Braille reader.

January 2001

CURRICULUM VITAE

Alfred H. Bellows Wayland, MA 01778
508 555-2424

EDUCATION:	M.S. Mechanical Engineering MIT, Cambridge, Massachusetts	1964
	B.S. Mechanical Engineering MIT, Cambridge, Massachusetts	1962
EXPERIENCE:	Staff Engineer Osram Sylvania, Osram Opto Semiconductors, Beverly, Massachusetts	1997 - 2005
	Senior Engineer, Principal Engineer, Staff Engineer GTE Laboratories Incorporated, Waltham, Massachusetts	1978 - 1997
	Program Manager Foster-Miller Associates, Waltham, Massachusetts	1975 - 1978
	Chief Mechanical Engineer, Manager of Manufacture Electronics Corporation of America, Cambridge, Massachusetts	1971 - 1975
	Engineer, Senior Engineer, Principal Engineer Polaroid Corporation, Cambridge, Massachusetts	1964 - 1971
	Research Assistant MIT, Cambridge, Massachusetts	1962 - 1964
CONSULTING:	Kluger, Peretz, Kaplan & Berlin (Digital camera evaluation)	2005 - 2006
	Osram Sylvania (Mold design for ceramic arc tubes)	2005
	Venture Scientific (Data acquisition and telemetry: offshore fuel cell)	2003
	Johnston Center for Sight (Development of eye surgery instrument)	2002 - 2006
	Greenblum & Bernstein (Expert witness: patent infringement)	2002 - 2004
	Cellular Science (Vault environment and intrusion monitor)	2001 - 2002
	GeoSystems Analysis (Long-term temperature recorder for bore-holes)	2000 - 2004
	University of Pittsburgh (Data collector for a breathing monitor)	2000
	MRH Foundation (Motion, temp, and BP monitor for roving baboons)	2000
	Eastern Oceanics (Motor controller for sea-bottom probe)	1999
	Stroock & Stroock & Lavan (Expert witness: patent infringement)	1998 - 2006
	Virginia Institute of Marine Science (Microcomputer program)	1998
	Foster-Miller Inc. (Methods for cleaning nuclear reactor exchanger)	1998
	NASA-Lewis Research Center (Resistance measuring device)	1995 - 1997
	Case Western Reserve University (Crystal growth ampoule)	1993 - 1994
	Avant Incorporated (Cameras and shutters)	1980 - 1989
	JB Systems (Hardware for optical scanner)	1979 - 1981
	Norton Associates (Photographic equipment)	1971 - 1976
	Massachusetts General Hospital (Redesign optical instrument)	1970
	Icon Corporation (Turret design for automated lathe)	1967

CURRICULUM VITAE (Cont'd.)   **Alfred H. Bellows**

PATENTS: 33 patents issued
- 5 - Telephone equipment, fiber optics, video cable power
- 4 - Electrical power control equipment
- 1 - Underground mining equipment
- 1 - Utility burner flame control
- 8 - Cameras, related mechanisms, hardware, or accessories
- 8 - Photographic optics
- 1 - Photographic electronics
- 5 - other mechanisms, heat extractor, superconductor, head-lamp, Hg dispenser

PUBLICATIONS:
Over a dozen journal articles and presentations on solar collector optics, arc studies, and material processing aboard the Space Shuttle. Dozens of internal company publications, proposals, and reports.

LEGAL WORK:
1980 - Six days of oral deposition for the patent infringement case of Polaroid Corporation v. Eastman Kodak Company, Civil Action No. 76-1634, United States District Court, District of Massachusetts.

1995 - Expert Witness: Provided reconstructive analysis of conditions that resulted in an electrical injury; prepared a written report and 2 responses to rebuttal in the case of Timothy W. Leader and Beth A. Leader v. Metropolitan Edison v. GTE Corporation, Docket No. 93-SU-04855-01, Court of Common Pleas of York County, Pennsylvania.

1998 – 2006 Expert Witness: Reviewed and analyzed single-use cameras for patent infringements, prepared in-depth reports, witness statements, and rebuttal reports, and provided deposition and testimony In The Matter of Certain Lens-Fitted Film Packages, Investigation No. 337-TA-406, International Trade Commission, Washington, DC; including an initial investigation, two enforcement proceedings and a remand proceeding; continuation of similar issues before New York and New Jersey venues of civil actions.

1999 - Deposition regarding inventorship and patent ownership in the case of GTE Laboratories Incorporated v. Walter J. Beriont, Civil Action No. 98-5418, Middlesex Superior Court, Massachusetts.

2002 through 2004 - Expert Witness: Reviewed and analyzed point-and-shoot zoom cameras for patent infringements in re Asahi vs. Samsung, Civil Action No. 00-4893 (NHP), US District Court, New Jersey.

July 2006